Studies in Fuzziness and Soft Computing

Volume 334

Series editor

Janusz Kacprzyk, Polish Academy of Sciences, Warsaw, Poland
e-mail: kacprzyk@ibspan.waw.pl

About this Series

The series "Studies in Fuzziness and Soft Computing" contains publications on various topics in the area of soft computing, which include fuzzy sets, rough sets, neural networks, evolutionary computation, probabilistic and evidential reasoning, multi-valued logic, and related fields. The publications within "Studies in Fuzziness and Soft Computing" are primarily monographs and edited volumes. They cover significant recent developments in the field, both of a foundational and applicable character. An important feature of the series is its short publication time and world-wide distribution. This permits a rapid and broad dissemination of research results.

More information about this series at http://www.springer.com/series/2941

Pedro Ponce-Cruz · Arturo Molina
Brian MacCleery

Fuzzy Logic Type 1 and Type 2 Based on LabVIEW™ FPGA

Pedro Ponce-Cruz
Tecnologico de Monterrey
Campus Ciudad de México
Tlalpan, Distrito Federal
Mexico

Brian MacCleery
National Instruments Corporation
Austin, TX
USA

Arturo Molina
Tecnologico de Monterrey
Campus Ciudad de México
Tlalpan, Distrito Federal
Mexico

The Fuzzy Logic Type 1 and Type 2 Based on LabVIEW FPGA Toolkit can be downloading from the additional material of the book. For decompressing the toolkit you have to use the password (TOOLKITFPGA@TEC01). On the other hand, it is allowed to use the toolkit in academic and research implementations based on LabVIEW FPGAs but it has to be properly referenced. It is not allowed to use it in industrial applications without permission from the authors.

ISSN 1434-9922 ISSN 1860-0808 (electronic)
Studies in Fuzziness and Soft Computing
ISBN 978-3-319-26655-8 ISBN 978-3-319-26656-5 (eBook)
DOI 10.1007/978-3-319-26656-5

Library of Congress Control Number: 2015955366

Springer Cham Heidelberg New York Dordrecht London
© Springer International Publishing Switzerland 2016
This work is subject to copyright. All rights are reserved by the Publisher, whether the whole or part of the material is concerned, specifically the rights of translation, reprinting, reuse of illustrations, recitation, broadcasting, reproduction on microfilms or in any other physical way, and transmission or information storage and retrieval, electronic adaptation, computer software, or by similar or dissimilar methodology now known or hereafter developed.
The use of general descriptive names, registered names, trademarks, service marks, etc. in this publication does not imply, even in the absence of a specific statement, that such names are exempt from the relevant protective laws and regulations and therefore free for general use.
The publisher, the authors and the editors are safe to assume that the advice and information in this book are believed to be true and accurate at the date of publication. Neither the publisher nor the authors or the editors give a warranty, express or implied, with respect to the material contained herein or for any errors or omissions that may have been made.
LabVIEW™ is a trademark of National Instruments Corporation, 11500 N Mopac Expwy, Austin, TX 78759-3504, USA, http://www.ni.com/.

Printed on acid-free paper

Springer International Publishing AG Switzerland is part of Springer Science+Business Media (www.springer.com)

To Norma, Jamie, Pedro, Alize, Jorge, Aura, and Giovanna. Also, I dedicate this book to my lovely mother (Margarita) and grandmother (Catalina Vazquez†) who help me to dream in fuzzy colors

Pedro Ponce-Cruz

To my father Arturo, in memory of my mother Rosita and my family, Silvia my lovely wife and beloved kids Julio and Monse

Arturo Molina

For Eva Jane MacCleery. Most things in life and science have a fuzzy type of logic. However, if we listen carefully and are very fortunate, we find love, a mysterious clarity to guide us through this fuzzy world

Brian MacCleery

Foreword

Fuzzy logic is widely applied in every aspect of our daily lives. Washing machines, air conditioners, and vehicles are examples of applications in which fuzzy controllers are embedded to achieve smooth, nonlinear, and robust control. Such applications depend on a system architecture that is easy to implement, rather than on a theoretically abstract and complex methodology. Therefore, several tools, such as the NI LabVIEW PID and Fuzzy Logic Toolkit for Windows, have been proposed to assist developers in implementing fuzzy logic systems. Some tools for this purpose can even be downloaded for free and come with a user manual. Through the use of such tools, several successful applications have been demonstrated, which has further motivated the application of fuzzy logic in practice. However, most existing fuzzy logic applications in practice use ordinal fuzzy sets. Other fuzzy set types, such as interval-valued fuzzy sets, type-2 fuzzy sets, and hesitant fuzzy sets, have seldom been used. This is mainly because of the difficulties associated with implementing these generalized forms of fuzzy sets. However, several attempts have indicated that using these generalized fuzzy sets has several advantages. For example, the uncertainty behind a phenomenon can be more effectively described using a type-2 fuzzy set.

The authors of this book combined the latest research findings and practical experience with LabVIEW™ FPGA. I particularly appreciate the perspective on FPGA's most recent development strategy. This book also provides information for engineers seeking to understand fuzzy logic and how it can be applied to their products or system designs.

Prof. Dr. Tin-Chih Toly Chen
Outstanding Professor
Feng Chia University
Founding Editor-in-Chief
International Journal of Fuzzy System Applications

Preface

This book presents fuzzy logic and LabVIEW FGPA for designing fuzzy logic controllers. This is a book for implementing fuzzy logic controllers in LabVIEW FPGAs.

Despite the FPGA's attractive features, their adoption by industrial control and signal processing engineers has been slower than processors and DSPs. This is due to several factors. First, these engineers traditionally programmed processors and DSPs using higher level languages, such as C. However, FPGAs possessed complex development tool chains that required designs to be specified using hardware description level (HDL) and register transfer level (RTL) semantics. Furthermore, traditional FPGA development tools lacked intellectual property (IP) blocks for common industrial applications, such as ADC and encoder interface logic, PWM and commutation logic, timing and triggering functions, PID control algorithms, memory management, and data transfer functions. In addition, FPGAs natively supported integer data types only which significantly increased development complexity for analog control and signal processing applications that required math, control, and digital signal processing algorithms, as opposed to floating point processors. Also, traditional FPGA simulation tools were operated at the digital design level and were not interoperable with the type of dynamic simulation tools used by control systems and signal processing engineers for modeling continuous time dynamic system response. Moreover, FPGAs compilation times were relatively long, as compared to processors and DSPs. For example, typical FPGA compilation times today range from 15 to 90 min, whereas processor and DSP compilations are typically completed in less than one minute. Finally, the sequential text-based semantics of traditional register level development tools made it relatively difficult to specify timing and concurrency among parallel processing tasks in a way that leverages the inherent parallel processing capability of FPGA devices.

Despite these traditional development challenges, the successful adoption of FPGAs in application areas such as consumer electronics, and the resulting drop in the price of FPGAs has spurred the interest of industrial control design and simulation vendors. Such vendors are creating the next generation FPGA development tools that are designed for engineers with little or no digital design expertise. The goal of these next generation "system-level" graphical design tools is to empower control, simulation, and signal processing engineers to harness the full power of the FPGA technology. Graphical system design tools are intended to provide a more intuitive, high level programming paradigm that simplifies the creation of complex parallel processing and control applications. Also, they are intended to provide relatively competitive performance and resource usage, as compared to traditional HDL development tools.

Graphical data flow programming languages are a natural fit for FPGA development due to their inherent sense of parallelism and concurrency that naturally maps to hardware design. Also, recent technological advances are enabling designers to place their FPGA code within a high-level dynamic simulation environment. This ability to cross the boundary between the digital domain of the FPGA and the analog multi-physics domain of the system is facilitating a "true" mechatronics approach to development, in which the complex interplay between FPGA silicon logic, power electronic components, electric motor drives, and mechanical systems can all be simulated in a virtual environment without the need to wait for long FPGA compilations.

The ability to quickly iterate and optimize the FPGA logic design in a mechatronics simulation environment, combined with the new high-level programming tools for FPGAs is reducing dramatically the barriers that prohibited wide adoption of FPGAs in industrial control.

In addition to the improved design and simulation tools for FPGAs, the next generation tools are providing a rapidly growing library of IP blocks for common control and DSP algorithms through code sharing services. On the other hand, the number of books that present Fuzzy logic Control is big as Fuzzy logic control is one of the most important control techniques. However, several books are only mathematical descriptions and are not focused on implementation of fuzzy logic control. Moreover, there are not enough books that deal with implementing fuzzy logic controllers in FPGAs. There is still a lot of vagueness and misunderstanding around the implementation of fuzzy logic controllers implemented in LabVIEW™ FPGA.

Since this book presents a clear description of fuzzy logic control type 1 and 2 that are the most used fuzzy logic representations, the implementation in LabVIEW™ FPGA can be developed. Several experimental examples are presented in order to show the potential of Fuzzy Logic controllers implemented in FPGA.

Finally, a complete LabVIEW™ FPGA toolkit for fuzzy logic type 1 and type 2 is included in the book. This toolkit is based on fix point representation that LabVIEW™ FPGA needs. This toolkit is developed for working on LabVIEW™ real-time systems.

Pedro Ponce-Cruz
Arturo Molina
Brian MacCleery

Acknowledgments

We hereby acknowledge the following organizations for their contributions to this book: Tecnologico de Monterrey, Campus Ciudad de México and National Instruments, Austin, Texas.

Contents

Chapter 1
Literature Review for Digital Implementations of Fuzzy Logic Type-1 and Type-2

1.1 Advances in Applications of Fuzzy Logic Systems

Some works are described below where optimization Type-1 and Type-2 FLS have had relative success according to different areas, illustrating the advantages of using methods to automate process with fuzzy controllers.

The work of Sudha and Vijai Santhi [1], deals with the load frequency control (LFC) problem, which has been a major subject in electrical power system such as design and operation. In practice, LFC systems use simple proportional integral (PI) controllers. However, it shows that fuzzy logic controller is adaptive in nature and is applied successfully for power system stabilization control. Authors propose a type-2 fuzzy approach for load frequency control of two-area interconnected reheat thermal power system with the consideration of Generation Rate Constraint (GRC). The performance of the type-2 controller is compared with conventional controller and Type-1 fuzzy controller with regard to Generation Rate Constraint (GRC). The system parametric uncertainties are verified by changing parameters by 40% simultaneously from their typical values.

The authors Jammeth and Fleury in their work [2] use an interval Type-2 FLC that gets better video quality compared with traditional controllers. They also use Type-1 FLC to show the response in different network scenarios, such like Internet cross-traffic, video streams occupy on an Internet protocol (IP) network. Type-2 FLC is designed for Internet conditions so they can react successfully to the network conditions of an IP-network. Type-2 FLC resulted in an order of magnitude performance improvement in comparison with Type-1 FLC.

The work of Lee and Lin [3], proposes a new control scheme using type-2 fuzzy neural network (type-2 FNN) and adaptive filter for controlling nonlinear uncertain systems. This type-2 FNN model combines the advantages of type-2 fuzzy logic systems and neural networks. The type-2 FNN system has the ability of universal approximation, which identifies nonlinear dynamic systems. The control scheme consists of a PD-type adaptive FNN controller and a pre-filter. The adaptive filter is

© Springer International Publishing Switzerland 2016

P. Ponce-Cruz et al., *Fuzzy Logic Type 1 and Type 2 Based on LabVIEW™ FPGA*, Studies in Fuzziness and Soft Computing,
DOI 10.1007/978-3-319-26656-5_1

used to provide better performance under transient response and to treat the problem of disturbance attenuation. The tuning parameters for the filter and the type-2 FNN controller will change according to the learning algorithm. By the Lyapunov stability theorem, the convergence of parameters is given in order to guarantee the stability of nonlinear uncertain systems.

The work of Biglarbegian et al. [4], present a novel design methodology of interval Type-2 Takagi Sugeno Kang fuzzy logic controllers for modular and reconfigurable robot manipulators with uncertain dynamic parameters. Results show that the developed controller can perform better than some linear and non-linear controllers for different configurations. Then, the structure can be adopted for the position control of robots with unknown dynamic parameters in trajectory-tracking applications.

The work of Koca et al. [5], describes a new control scheme for the robust crank angular speed control of a four-bar mechanism driven by a DC motor, based on type-2 fuzzy logic and sliding mode control (SMC) technique. Type-2 fuzzy logic and SMC can be combined to use the advantages of both methods and thus to improve the effectiveness of the controllers. One of the most important advantages of the use of SMC with type-2 FC is to reduce the number of fuzzy rules and to obtain a simpler and more practical control algorithm to use in real applications.

Poornaselvan et al. [6], focus on an agent-based approach to flight control in ground. A type-2 fuzzy interval controller can be applied to the autonomous vehicle in order to handle uncertainty in a better way. Ant colony optimization technique can be used to optimize path planning in traffic environment. They use a hybrid ant colony optimization to handle real-time dynamic environment and path planning. Both agents-based and type-2 fuzzy together with ant colony optimization technique are used to achieve another level of intelligence.

Zaher and Hagras [7], present a method to generate a type-2 FL model entirely from data to provide a dynamic footprint of uncertainty for the generated fuzzy set. The fuzzy model will be used to predict the wind speed experienced by a wind turbine without the use of sensors. This estimated wind speed is then passed for another fuzzy controller that changes the pitch angles of the wind turbine blades in order to track the maximum power available.

In the work of Galluzzo and Cosenza [8], they develop two adaptive type-2 FL controllers with minimum number of rules, compared by simulation for control of a bioreactor in which aerobic alcoholic fermentation for the growth of *Saccharomyces cerevisiae* takes place. The bioreactor model is characterized by nonlinearity and parameter uncertainty. The first adaptive fuzzy controller is a type-2 fuzzy-neuro-predictive controller (T2FNPC) that combines the capability of type-2 FL to handle uncertainties, with the ability of predictive control to predict future plant performance making use of a neural network model of the nonlinear system. The second adaptive fuzzy controller is instead a self-tuning type-2 PI controller, where the output scaling factor is adjusted online by fuzzy rules according to the current trend of the controlled process.

The work of Chaoui and Gueaieb [9], a type-2 FL controller is applied to robot manipulators with joint elasticity and structured and unstructured dynamical uncertainties. The proposed controller is based on a SMC strategy. To enhance its real-time performance, simplified interval fuzzy sets are used. The efficiency of the control scheme is further enhanced by using computationally inexpensive input signals independent of the noisy torque and acceleration signals, and by adopting a trade-off strategy between the manipulator's position and the actuators' internal stability. The controller is validated through a set of numerical experiments and by comparing it against its type-1 counterpart. It is shown through these experiments, the higher performance of the type-2 FLC in compensating for larger magnitudes of uncertainties with severe nonlinearities.

The work of Petrović et al. [10], deals with the systematic maintenance of mining machinery and equipment. They present a model of the risk assessment of technical systems failure based on the fuzzy sets theory, fuzzy logic, and min–max composition. The risk indicators, severity, occurrence, and detectability are analyzed. The risk indicators are given as linguistic variables. The model was applied for assessing the risk level of belt conveyor elements failure which works in severe conditions in a coal mine.

Baldania et al. [11], focuses on rule base division technique to reduce the power consumption of the fuzzy logic controller based on different conditions and priorities. A new methodology is developed and implemented successfully to reduce the power consumption of the fuzzy logic controller, i.e., dynamic rule base method. The purpose is to solve complex problems efficiently with better accuracy and fast time being response. Fuzzy logic controller has the potential to solve problems which have uncertain nature and complex behavior with better, accurate, and efficient results and also with faster time response as compared to classical logic-based general purpose controllers or processors like PID controllers. The development of the fuzzy logic-based controller follows a step-by-step approach with expertize knowledge. However, there is a lack of power reduction techniques for fuzzy logic controller when the application is concerned for embedded applications. As embedded devices are battery operated and have limited power supply, there is a need of a proper approach and method to reduce the power consumption.

The work of Novak et al. [12], presents three special soft computing software systems, based on the original results in two areas: fuzzy natural logic and fuzzy transform. The first software is LFL Controller which is a universal system that can be used in fuzzy or linguistic control, and in decision making. The system implements results of fuzzy natural logic, namely the theory of evaluative linguistic expressions and perception-based logical deduction. The second system is LFL Forecaster which is a specialized SW for analysis and forecasting of time series. The analysis is realized using F-transform and forecasting using results of fuzzy natural logic. The third system is FT-Studio which is specialized for computation of fuzzy transform of functions that can be defined either using a formula, or given by data.

In the work of Arbex et al. [13], presents a model which uses fuzzy logic as the basis for the development of an inference system to aid decision making, which is based on previous results obtained by different single nucleotide polymorphisms (SNP) discovery tools and which present possibly conflicting results. A single difference between base pairs of different aligned sequences is the most common type of genetic variability and they are called single nucleotide polymorphisms (SNPs). They are crucial for the study of species variability, since they can cause functional or phenotypic modifications which, in turn, may imply in evolutionary or biochemical consequences in the individuals of a given species.

Ai-Zhen and Feng [14], made a fuzzy controller neural network for a washing machine to reduce the wasting of electric and water. The washing machine fuzzy controller neural network is researched deeply, which is based on fuzzy logic, neural network, and its learning algorithm. The BP neural network is combined with fuzzy control and experiments are simulated by MATLAB. Water level, flow intensity, and the washing time are preset. Fuzzy control rules and membership functions are automatically generated. These parameters can be adjusted real-time to improve the performance of washing machines and achieve better water-saving effect of energy saving.

The work of Liang and Mendel [15], presents a type-2 fuzzy adaptive filter (FAF); one that is realized using an unnormalized type-2 Takagi-Sugeno-Kang (TSK) fuzzy logic system (FLS). They apply this filter to equalization of a nonlinear time-varying channel and demonstrate that it can implement the Bayesian equalizer for such a channel, has a simple structure, and provides fast inference. In 2001 [16], they presented an approach for MPEG variable bit rate (VBR) video modeling and classification using fuzzy techniques. They demonstrated that a type-2 fuzzy membership function, i.e., a Gaussian MF with uncertain variance, is most appropriate to model the log-value of I/P/B frame sizes in MPEG VBR video. The fuzzy c-means (FCM) method is used to obtain the mean and standard deviation (std) of T/P/B frame sizes when the frame category is unknown. They proposed to use type-2 fuzzy logic classifiers (FLCs) to classify video traffic using compressed data. Five fuzzy classifiers and a Bayesian classifier are designed for video traffic classification, and the fuzzy classifiers are compared against the Bayesian classifier. Simulation results show that a type-2 fuzzy classifier in which the input is modeled as a type-2 fuzzy set and antecedent membership functions are modeled as type-2 fuzzy sets performs the best of the five classifiers when the testing video product is not included in the training products and a steepest descent algorithm is used to tune its parameters.

The work of Mitchell [17], introduces a similarity measure for measuring the similarity, or compatibility, between two type-II fuzzy sets. With this new similarity measure, he shows that type-II fuzzy sets provide us with a natural language for formulating classification problems in pattern recognition. Type-II fuzzy sets are a generalization of the ordinary fuzzy sets in which the membership value for each member of the set is itself a fuzzy set in [0, 1].

The work of Herman et al. [18], examines the potential of the type-2 FLS methodology in devising an EEG-based brain–computer interface (BCI). In particular, a type-2 FLS has been designed to classify imaginary left- and right-hand movements based on time-frequency information extracted from the EEG with the short-time Fourier transform (STFT). The type-2 fuzzy classifier has been proven to outperform its type-1 counterpart on all datasets recorded from three subjects examined. It has also compared favorably to the well-known classifier based on linear discriminant analysis (LDA). Analysis of the electroencephalogram (EEG) requires a framework that facilitates handling the uncertainties associated with the varying brain dynamics and the presence of noise. Recently, the type-2 fuzzy logic systems have been found effective in modeling uncertain data.

In the work of Karnik and Mendel [19], they begin with a type-1 fuzzy logic system (FLS), trained with noisy data. They demonstrate how information about the noise in the training data can be incorporated into a type-2 FLS, which can be used to obtain bounds within which the true (noise-free) output is likely to lie. They do that with the example of a one-step predictor for the Mackey–Glass chaotic time series. They also demonstrate how a type-2 FLS can be used to obtain better predictions than those obtained with a type-1 FLS.

The paper of Ozen and Garibaldi [20], explains how the shape of type-2 fuzzy membership functions can be used to model the variation in human decision making. An interval type-2 fuzzy logic system (FLS) is developed for umbilical acid–base assessment. The influence of the shape of the membership functions on the variation in decision making of the fuzzy logic system is studied using the interval outputs. Three different methods are used to create interval type-2 membership functions. The center points of the primary membership functions are shifted, the widths are shifted, and a uniform band is introduced around the original type-1 membership functions. It is shown that there is a direct relationship between the variation in decision making and the uncertainty introduced to the membership functions.

The work of Liang and Wang [21], presents a new approach for sensed signal strength forecasting in wireless sensors using interval type-2 fuzzy logic system (FLS). They show that a type-2 fuzzy membership function with uncertain mean is most appropriate to model the sensed signal strength of wireless sensors. They demonstrate that the sensed signals of wireless sensors are self-similar, which means they can be forecasted. An interval type-2 FLS is designed for sensed signal forecasting and is compared against a type-1 FLS. Simulation results show that the interval type-2 FLS performs much better than the type-1 FLS in sensed signal forecasting. This application can be further used for power on/off control in wireless sensors to save battery energy.

The work of Castillo and Melin [22], describes the application of type-2 fuzzy logic for achieving adaptive noise cancelation. The objective of adaptive noise cancelation is to filter out an interference component by identifying a model between a measurable noise source and the corresponding unmeasurable interference. They propose the use of type-2 fuzzy logic to find this model. The use of type-2 fuzzy logic is justified due to the high level of uncertainty of the process, which makes it difficult to find appropriate parameter values for the membership functions.

In the work of Lee et al. [23], presents a type-2 fuzzy neural network system (type-2 FNN) and its learning algorithm using backpropagation algorithm. The FNN system using type-1 fuzzy logic systems (FLS) is called type-1 FNN system. It has the properties of parallel computation scheme, easy to implement, fuzzy logic inference system, and parameters convergence. For considering the fuzzy rules uncertainties, we use the type-2 FLSs to develop a type-2 FNN system. The type-2 fuzzy sets let to model and to minimize the effects of uncertainties in rule-based fuzzy logic systems (FLSs). The previous results of type-1 FNN are extended to a type-2 one. In addition, the corresponding learning algorithm is derived by back-program algorithm. Several examples are presented to illustrate the effectiveness of the model.

The work of Rhee and Hwang [24], presents an interval type-2 fuzzy perceptron algorithm that is an extension of the type-1 fuzzy perceptron algorithm proposed by Keller et al. [25]. In their proposed method, the membership values for each pattern vector are extended as interval type-2 fuzzy memberships by assigning uncertainty to the type-1 memberships. By doing so, the decision boundary obtained by interval type-2 fuzzy memberships can converge to a more desirable location than the boundary obtained by crisp and type-1 fuzzy perception methods.

The work of Agero and Vargas [26], presents an approach of aiming at inferring the operative configuration (OC) of distribution networks. In order to attain this objective, the OC problem is decomposed in two parts. The first part, the available real-time data and expert knowledge are integrated by means of rule-based type-2 FLS. As a result, an approximate initial solution is obtained. In the second part, the initial solution is dynamically tuned using customer trouble calls. This is done by means of an approach based on fuzzy relational equations and fuzzy inference. The performance of the methodology is evaluated on a real distribution feeder and the results are presented.

1.2 FPGA and Microcontrollers Used for Fuzzy Logic Applications

Field programmable gate array (FPGA) are two-dimensional arrays of logic blocks and flip-flops with an electrically programmable interconnection between logic blocks. The interconnections consist of electrically programmable switches which is why FPGA differs from custom integrated circuits, as a custom integrated circuit is programmed using the technology to form metal interconnections between logic blocks. In an FPGA, logic blocks are implemented using multiple-level low fan in gates, which gives it a more compact design compared to an implementation with two-level AND-OR logic [27]. Microcontrollers and FPGA’s provide functional blocks to build type-1 FLS and type-2 FLS, in which crisp inputs and outputs, and parameters are defined in 8-bits. It means that in a byte one can spread all inputs or outputs discourse universe. A crisp input may be defined in speed, pressure, temperature, or another physical variable. When an inferred value is ready, the next step is to convert it into a physical value so it can be used as the control result applicable to the plant of the controller. Some applications are described since microcontrollers, DSP up to FPGA.

1.2.1 Microcontroller Application

The work of Muscato [28], the position control of a planar under actuated manipulator with two revolute joints is considered. A dynamic model of the system is presented and a fuzzy control strategy is proposed. Fuzzy logic allows empirical rules to be translated into a control algorithm. A fuzzy microcontroller is adopted for the practical implementation of the system. The results of several experiments are presented and discussed.

The work of Faravelli et al. [29], pursues the numerical testing of a programmable controller driving an active mass damper (AMD) located at the top of a frame structure subject to base excitation. Its key features are the fuzzy nature of the controller and its adaptive characteristics toward fault tolerance. Numerical testing of microcontrollers is a reasonable way to design their parameters in a consistent environment. Indeed, in such a way, one avoids unrepairable failures of the device to be controlled and the consequent costs. A final experimental test will eventually provide the fine tuning of the parameters.

In the study of El-Nagar and El-Bardini [30], they propose an embedded real-time interval type-2 fuzzy proportional–integral–derivative (IT2F-PID) controller which is a parallel combination of the interval type-2 fuzzy proportional–integral (IT2F-PI) controller and the interval type-2 fuzzy proportional–derivative (IT2F-PD) controller. The proposed IT2F-PID controller is able to handle the effect of the system uncertainties due to the structure of the interval type-2 fuzzy logic controller. The proposed IT2F-PID controller is implemented practically using a low cost PIC microcontroller for controlling the uncertain nonlinear inverted pendulum to minimize the effect of the system uncertainties due to the uncertainty in the mass of the pendulum, the measurement error in the rotation angle of the pendulum and the structural uncertainty. The test is carried out using the hardware-in-the-loop (HIL) simulation. The experimental results show that the performance of the IT2F-PID controller improves significantly the performance over a wide range of system uncertainties.

1.2.2 DSP Application

The work of Bal et al. [31], develops a digitally controllable two phase serial resonant inverter to drive the ultrasonic motor by using a TMS320F243 digital signal processor. The driving frequency was used as a control input in the position control loop. The position characteristics obtained from the proposed drive and control system were demonstrated and evaluated by experiments. The experimental results verify that the developed position control scheme is highly effective, reliable and applicable for the ultrasonic motor. Position control of an ultrasonic motor was implemented on the basis of fuzzy reasoning.

In the work of Goh et al. [32], they improve the overall performance of an overcurrent relay for power system protection compared to conventional relay. It is essential for a relay to work efficiently to trip the circuit breakers in the presence of faults and at the same time proficient to coordinate well with the networks to avoid maloperation. There are two different types of fuzzy logic control strategies proposed for the relay, the Fuzzy Logic Controller (FLC) and Fuzzy Bang-Bang Controller (FBBC). The FBBC is the same as the conventional FLC except that the defuzzification method uses largest of maxima (LOM). Comparisons between the fuzzy controllers and conventional relay are based on IEC 255-3 standard. These relays are implemented on a DSP TMS320F2812 and their performance is evaluated which is based on operation time, DSPs execution time and grading margin. The results obtained show a significant performance improvement compared to conventional relay.

Goh et al. as well [33], improve the performance of the relay significantly with the new numerical relay technology using digital signal processor (DSP). However, application of DSP in numerical overcurrent relays is limited especially in coordination among the group of relays. The relay must work proficiently to coordinate with the networks in order to avoid maloperation. Therefore, in this paper, an implementation of overcurrent relay with improved coordination on a DSP, TMS320F2812 is described. The fuzzy bang–bang controller is used as the control strategy for the relay to provide efficient control for overcurrent protection. The performance evaluation of the proposed system is based on steady-state analysis, transient state analysis, coordination and lastly the execution time of the DSP. The results obtained using this new proposed controller is very promising. Overcurrent relays are very important protection components that require high reliability to maintain high security in power systems.

The study of Chou et al. [34], proposes a digital signal processor (DSP)-based complementary sliding mode control (CSMC) with Sugeno type fuzzy neural network (SFNN) compensator for the synchronous control of a dual linear motors servo system installed in a gantry position stage. The dual linear motors servo system comprises two parallel permanent magnet linear synchronous motors (PMLSMs). The dynamics of the single-axis motion system with a lumped uncertainty which contains parameter variations, external disturbances, and nonlinear friction force are briefly introduced first. Then, a CSMC is designed to guarantee the precision position tracking requirement in single-axis control for the dual linear motors. Moreover, to enhance the robustness to uncertainties and to eliminate the synchronous error of dual linear motors, the CSMC with a SFNN compensator is proposed where the SFNN compensator is designed mainly to compensate the synchronous error. Furthermore, to increase the control performance of the proposed intelligent control approach, a 32-bit floating-point DSP, TMS320VC33, is adopted for the implementation of the proposed CSMC and SFNN.

1.2.3 FPGA Application

The work of Melgarejo and Peña-Reyes [35], presents an architectural proposal for a hardware-based interval type-2 fuzzy inference system. First, it presents a computational model which considers parallel inference processing and type reduction based on computing inner and outer bound sets. They infer that a hardware architecture with several pipeline, stages for full parallel execution of type-2 fuzzy inferences. The architectural proposal is used for specifying a type-2 fuzzy processor with reconfigurable rule base, which is implemented over FPGA technology.

The review of state of the art of FPGA of Sulaiman et al. [27], approach a large numbers of fuzzy control applications with the physical systems required a real-time operation to interface high speed constraints; higher density programmable logic devices such as field programmable gate array (FPGA) can be used to integrate large amounts of logic in a single integrated circuit (IC). In their review, they focus on FPGA-based fuzzy logic controller. The paper starts with an overview of FPGA in order to get an idea about FPGA architecture, and followed by an explanation on the hardware implementation with both type analog and digital implementation, a comparison between fuzzy and conventional controller are also provided. A survey on fuzzy logic controller structure is highlighted with the focus on FPGA-based design of fuzzy logic controller with different applications. Finally, they provide the simulation and experimental results from the literature and conclude the main differences between software-based systems with respect to FPGA-based systems, and the main features for FPGA technology and its real-time applications.

The work of Ramadan et al. [36], presents an improved adaptive fuzzy logic speed controller for a DC motor, based on FPGA hardware implementation. The developed controller includes an adaptive fuzzy logic control algorithm, which is designed and verified with a nonlinear model of DC motor. Then, it has been synthesized, functionally verified, and implemented using Xilinx Integrated Software Environment (ISE) and Spartan-3E FPGA. The performance of the controller has been successfully validated with good tracking results under different operating conditions.

The work of Messai et al. [37], details the work in which they optimize and implement a FLC used as a maximum-power-point tracker for a stand-alone PV system, are presented. The near optimum design for membership functions and control rules were found simultaneously by genetic algorithms which are search algorithms based on the mechanism of natural selection and genetics. These are easy to implement and efficient for multivariable optimization problems such as in fuzzy controller design. The FLC thus designed, as well as the components of the PV control unit, were implemented efficiently on a Xilinx reconfigurable FPGA chip using VHDL Hardware Description Language. The obtained simulation results confirm the good tracking efficiency and rapid response to changes in environmental parameters.

In the work of Soares dos Santos and Ferreira [38], tests if FPGAs are able to achieve better position tracking performance than software-based soft real-time platforms. For comparison purposes, the same controller design was implemented in these architectures. A Multi-state Fuzzy Logic controller was implemented both in a Xilinx® Virtex-II FPGA (XC2v1000) and in a soft real-time platform NI CompactRIO®-9002. The same sampling time was used. The comparative tests were conducted using a servo-pneumatic actuation system. Steady-state errors lower than 4 μm were reached for an arbitrary vertical positioning of a 6.2 kg mass when the controller was embedded into the FPGA platform. Performance gains up to 16 times in the steady-state error, up to 27 times in the overshoot, and up to 19.5 times in the settling time were achieved by using the FPGA-based controller over the software-based FLC controller.

The work of Munoz et al. [39], describes the validation of five dispatching algorithms for elevator systems that were implemented on Spartan 3 FPGA-based boards in an integrated approach reducing the area and improving performance. The overall system is composed of several local control systems (LCS), which implement the dispatching algorithms, an RS485-based network and a virtual environment called virtual elevator interface system (VEI), which includes a simulator/monitoring system and an elevator group control system EGCS-based on fuzzy logic (FEGCS). The FEGCS runs on a PC and, under different traffic situations, determines the best algorithm to be run in each LCS in order to reduce the user waiting time and the power consumption. The novelty of this approach is that the LCSs are capable to run different dispatching algorithms independently, that are suitable for specific passenger traffic situations, while the FEGCS only must determine the best algorithm to be run in each LCS. The VEI allows the designer to test and validate in a flexible way the algorithm performance for different traffic situations. Elevator systems are administrated by an EGCS and microprocessed subsystems implementing a LCS for each elevator.

1.3 Fuzzy Logic Concepts

Lofty Zadeh [40], a computer scientist at the University of California Berkeley, proposed the fuzzy set theory in 1965. "A fuzzy set is a class of objects with a continuum of grades of membership." Lofty defined fuzzy sets as a class of sets with grades of membership from 0 to 1. Expert Systems, Artificial Neural Networks and Fuzzy Systems share the property of being model-free approximations, which means that no exact mathematical model of the physical system to control or to approximate is needed.

One of the most important areas of implementing fuzzy logic is in control systems. Fuzzy logic control has been in several real-world applications. The first Fuzzy logic controller was developed by Mamdani and Assilian in 1975 for controlling a steam generator in a laboratory setting and Blue Circle Cement and SIRA in Denmark developed a cement kiln controller that is the first industrial application in 1976. Japan developed several industrial applications using fuzzy logic. For instance, a water treatment system developed by Fuji Electric. Besides, Hitachi installed a fuzzy logic-based automatic train operation control system into the Sendai city's subway system based on fuzzy logic. These applications of fuzzy logic attracted the attention of many engineers to use fuzzy logic in control systems.

Fuzzy sets intend to model the uncertainty or vagueness associated to the natural human reasoning, which is based on linguistic words and sentences rather than in mathematical expressions and relations. Approximate reasoning or fuzzy reasoning is a mode of reasoning which is neither exact nor inexact [Zade75]. It is supported on the Fuzzy Logic theory and it offers a realistic framework for representing human reasoning. Approximate reasoning is the fundament for Fuzzy Inference Systems. To understand how it is performed, three basic concepts should be defined:

- Linguistic Variable. A linguistic variable is a variable whose values are words or sentences in a natural or artificial language rather than numerical. For instance, the variable speed can be described as it is presented below.

$$\mathrm{TS(speed)} = \{\mathrm{small, medium, big}\} = \{\mathrm{S, M, B}\}.$$

- Fuzzy Proposition. Fuzzy proposition is a statement expressed in a natural or artificial language. In contrast to classical logic propositions, a fuzzy proposition may adopt a truth-value from the interval [0, 1].For example, Distance is Big.
- Linguistic Rule. A linguistic if-then rule has two parts:

  Antecedent part (premise), expressed by: if <fuzzy proposition>,
  Consequent part, expressed by: then <fuzzy proposition>,
  It is used for setting the actions that can be done in a controller. For instance, if the distance is big then the controlled speed is big.

A fuzzy set is characterized by a membership function which assigns to each element a grade of membership. A fuzzy set F is entirely defined by the set of ordered pairs.

$$F = \{(x, \mu_F(x)) | X \in U\} \quad \text{and} \quad \mu_F: U \rightarrow [0, 1],$$

where x is an element of the universe of discourse U and μ_F is a membership function that assigns a degree of membership $\mu_F(x)$ to each element x of F.

In 1996, Zadeh [41] expressed a novel fuzzy logic approximation that is equated to computing with words (CW). There are two main ideas for computing with words. Initially, CW is a necessity when the available information is imprecise, and there is a tolerance for imprecision which can be exploited to achieve tractability, robustness, low solution cost, and better rapport with reality.

Fuzzy logic gradually emerged as a discipline in Artificial Intelligence, as well as expert systems and neural network [42]. Fuzzy logic helps solve complex problems, and it has been applied to several applications [43] like process control [1–9, 11, 14, 44–47], modeling [48], identification [48, 49], medicine [18], science [8, 14, 42, 50–52], and so on. Fuzzy Logic Type-2 is much computationally intensive comparing to its type-1 counterpart. Despite their computation burden, type-2 has been applied to various type of application such as, neural network [24], signal processing [15, 16], pattern recognition [17, 18], time series forecasting [19], wireless communication [21] decision making [20], noise canceling [22], system identification [23], and power engineering [26].

Fuzzy logic deals with real problems that have imprecise information or uncertainty. Fuzzy logic is defined as the set of mathematical descriptions based on degrees of membership. A fuzzy model includes linguistic rules that help to map the inputs and outputs of the system. Inputs in a Fuzzy logic systems can get membership values between 0 and 1. In conventional logic, the membership value is limited to only two values, 0 or 1 [53]. A Type-1 FLS has a grade of membership that is crisp, whereas Type-2 FLS has grades of membership that are fuzzy. Type-2 FLS has four basic components: fuzzification, evaluation of rules (if-then linguistic rules) and defuzzification. When membership functions are used, they can describe real-world situations where it is difficult to use binary membership values (0 or 1).

Two sets are presented in Fig. 1.1, the one on the left side is crisp and an element defined within a universe can belong or not to the set (element "*a*" belong the set or has a membership value of 1 and elements "*b*" and "*c*" do not belong and have a membership value of 0). On the other hand, the fuzzy set has different degrees of membership. For example, the element *f* has a lower degree of membership that elements *d* and *e* (if the center of the set represents the maximum membership value). A fundamental concept in mathematics is the notion of set. A set is a collection of specific, discernible elements. A set can be finite countable or uncountable elements and it could be described in two forms:

By naming all its elements:

$$A = \{a, b, c, 10, X\} \Rightarrow a, b, \ldots X \in A; \quad \text{where as} f,\ \beta \notin A.$$

By stating a property for all its elements:

$$A = \{x|P(x)\}, \quad \text{where } P \text{ defines the properties of the elements } x \text{ belonging to } A\ (\text{i.e.} A = \{x|x\} > 3)$$

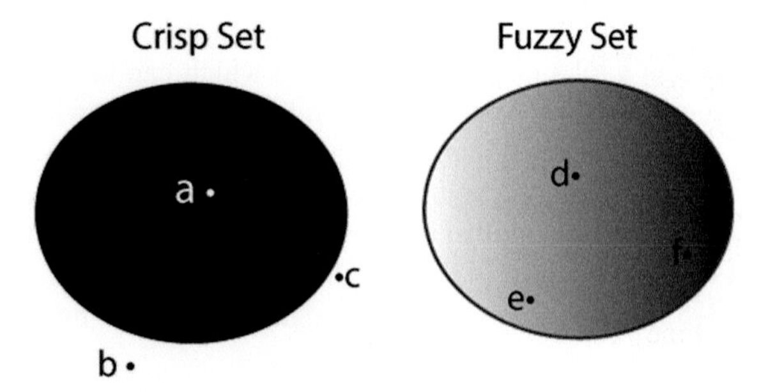

Fig. 1.1 Crisp set and fuzzy set

On the other hand, fuzzy sets can be represented by discrete or continues representation. A notation convention for fuzzy sets when the universe of discourse, X, is discrete and finite, it is as follows for a fuzzy set A:

$$\tilde{A} = \left\{ \frac{\mu_{\tilde{A}}(x_1)}{x_1} + \frac{\mu_{\tilde{A}}(x_2)}{x_2} + \cdots \right\} = \left\{ \sum_i \frac{\mu_{\tilde{A}}(x_i)}{x_i} \right\}$$

If the universe, X, is continuous and infinite, the fuzzy set A is defined by:

$$\tilde{A} = \left\{ \int \frac{\mu_{\tilde{A}}(x)}{x} \right\}$$

It can be defined as a *Characteristic Function* for all the elements x of the Universe of Discourse U. The set U is also called the Super Set. Therefore, for a set A belonging to U, the *Characteristic Function in a crisp set* (This function is called *membership function* in Fuzzy Set Theory) *can be defined by*

$$\mu_A(x) = \begin{cases} 1 & \text{if} \quad x \in A \\ 0 & \text{if} \quad x \notin A \end{cases}$$

The membership function of a fuzzy set corresponds to the characteristic function of a crisp set. However, while the characteristic function of a crisp set can only take values 0 or 1, the membership function of a fuzzy set can take any value from the interval [0, 1]. Hence, a fuzzy set is a generalization of the crisp set.

1.3.1 Type-1 Fuzzy Set (T1Fs)

Normally, Human decision thinking is not defined with mathematical methods or crisp numbers, so fuzzy numbers can be used for solving basic and advanced problems such as driving a car that deals with vague or ambiguous conditions. For example, when you are taking a shower and the temperature of the water is very hot, then your control action will be to close a little the hot water valve and open the cold water valve moderately; you can use labels like "little" and "moderately" to describe fuzzy values in the elements. Those elements have a membership value in a fuzzy set. For example, the label "a little" could mean a value between 10° and 30° degrees of rotation of the valve and "moderately " a rotation between 20° and 40° of rotation of the valve. Of course, we do not have a mathematical model which relates the temperature of the water with the rotation of the valves but we can achieve a comfortable temperature of the water without problems. Fuzzy set can help solve this description without knowing a mathematical model.

Fuzzy sets are used for describing fuzziness based on degrees of membership and they can be used in many real situations with linguistic terms: measurement of temperature, speed, distance, height, etc.; for instance: The air is **fresh**, the electric vehicle is **very fast** but it can only achieve **short** distance trips, Tony is a **very tall** guy. But considering these scales, very often it is difficult to distinguish between memberships of some class of nonmembers. One of the problems that can arise when fuzzy logic is used is how to determine the values for describing the fuzzy terms. For example, how long is a short distance trip? It is a question that can generate several linguistic responses. Fuzzy logic type-2 can deal with this kind of problems because it can deal with uncertainties in words of different persons.

On the other hand, crisp distinctions are used to model this with conventional or Boolean sets and it forces an abrupt transition between the members and not members of a set; for example, we may say that a short trip is less than 100 km (e.g., 99.9 km) and by deduction, a trip larger than 100.1 km is in a long trip. For avoiding these inconsistencies, fuzzy logic could be used.

1.3.2 Membership Function

When the Universal Set U is continuous or uncountable rather than discrete, membership functions can be expressed analytically by using continuous or piecewise explicit mathematical functions The basic idea of the Boolean logic is using only two membership values: true (1) or false (0), this Boolean logic cannot represent values between 1 and 0 that are used for defining with vague concepts. Let X be a Boolean (crisp) set and x an element of that set, then x belong to X $(x \in X)$ or x do not belong to X $(x \notin X)$. The classical theory imposes a sharp boundary of the sets and all the values that belong to x have a set value of 1 and the others which do not belong have a value of 0. However, crisp sets do not deal with membership values. The principle of a fuzzy set is based on membership values. For instance, information can be used for describing a table of maximum distances achieved by electric vehicles, and it is possible to assign a certain degree of membership to each element. A membership function called "long distance" can be used to describe the set distance (the description of the fuzzy set can be done by discrete or continues forms). For illustrative purposes, all the membership functions shown in the following figures will be continuous. Table 1.1 presents a description for variable long distance.

Using continues approximations, it is possible to get the following representation (fuzzy membership function) in fuzzy and crisp forms (see Figs. 1.2 and 1.3).

In Fig. 1.2 is shown a fuzzy set for "long distance;" the horizontal axis represents the universe of discourse in this case which is between 50 and 200 km and the vertical axis represents the degree of membership; for crisp sets, the transition between nonmembership value and membership value is abrupt; so a value of 99 will have a degree of membership of 0 while 100 has a value of 1. On the other hand, for a fuzzy set the transition could have the form of any function; in this case, a line was selected for simplicity but it can have any shape according to the suggestions of an expert or a knowledge database, more recently artificial neural

Table 1.1 Fuzzy set used for describing "long distance" in a discrete form

Vehicle	Distance	Crisp value	Fuzzy membership value
1	80	0	0.2
2	160	1	1
3	90	0	0.3
4	100	1	0.4
5	103	1	0.43
6	70	0	0.1
7	170	1	1
8	190	1	1
9	120	1	0.6
10	144	1	0.84

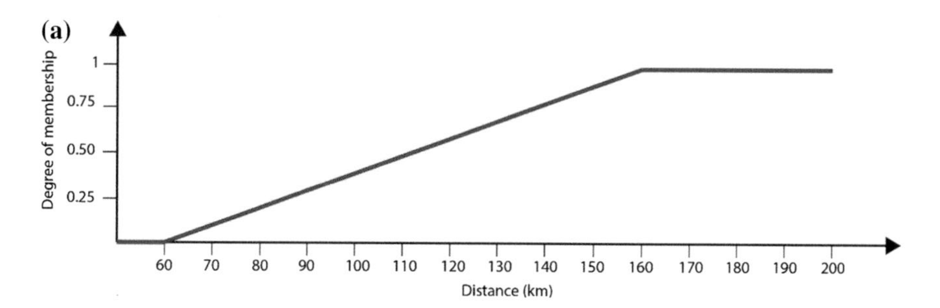

Fig. 1.2 Fuzzy set for "long distance"

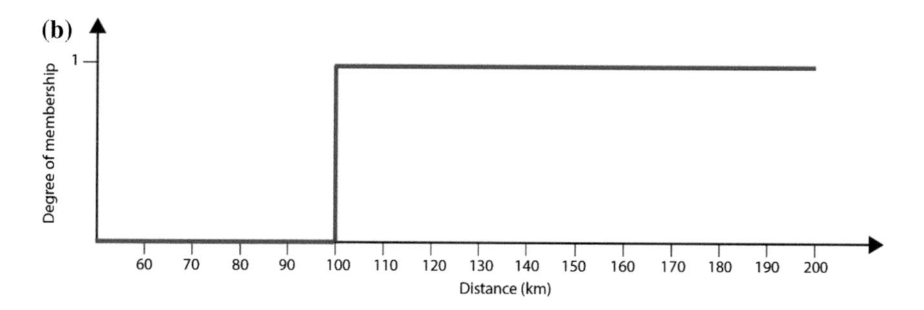

Fig. 1.3 Crisp set for "long distance"

networks are combined with fuzzy logic on order to learn from the available data and derive the value of the fuzzy sets automatically. In the next sections, the most employed membership function are presented and their mathematical representation. Figure 1.4 depicts conventional membership functions.

The main parts of the fuzzy membership function can be defined by the following terms (see Fig. 1.15):

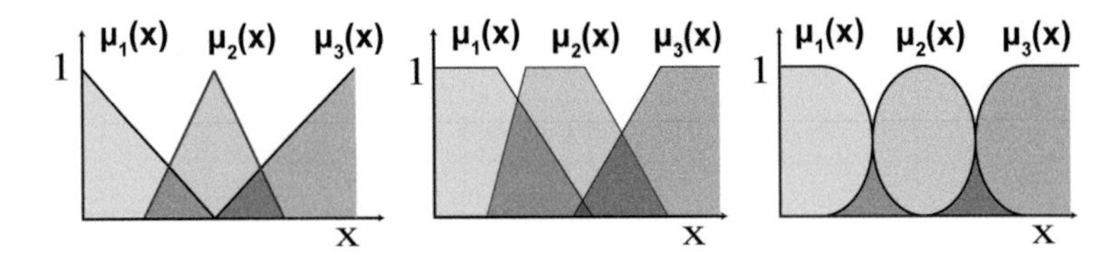

Fig. 1.4 Different shapes of conventional membership functions

The *core* of a membership function for some fuzzy set A, is defined as that region of the universe that is characterized by complete and full membership value $\mu_A(x) = 1$ in the set A. So, the elements in the set A that have a membership value equal to 1, are the elements that are part of the core.

The *support* of a membership function for some fuzzy set A is defined as that region of the universe that is characterized by nonzero membership value in the set A.

The *boundaries* of a membership function for some fuzzy set A are defined as that region of the universe containing elements that have a nonzero membership but not complete membership (Fig. 1.5).

As it was shown, there are several MF which can be used in T1FS; for example, the next normal membership functions presented above, A *normal or conventional* fuzzy set is one whose membership function has at least one element x in the universe whose membership value is one. The mathematical representation for some conventional membership functions is presented below.

Triangular MF

A triangular MF is defined by its parameters $[a, b, c]$ such

$$\mu_A(x) = \begin{cases} \frac{x-a}{b-a} & a \leq x \leq b \\ \frac{c-x}{c-b} & b \leq x \leq c \\ 0 & x \leq a \vee x \geq c \end{cases} \tag{1.1}$$

The shape of this function is shown in Fig. 1.6a.

Fig. 1.5 Membership functions parts

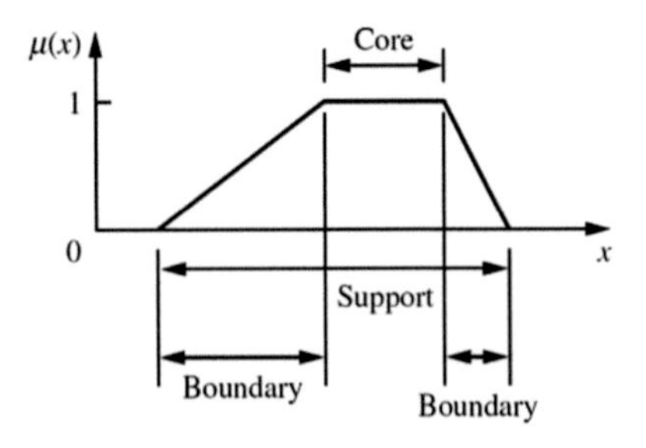

Trapezoidal MF

A trapezoidal MF is defined by its parameters $[a, b, c, d]$ such

$$\mu_A(x) = \begin{cases} \frac{x-a}{b-a} & a \leq x \leq b \\ 1 & b \leq x \leq c \\ \frac{d-x}{d-c} & c \leq x \leq d \\ 0 & x \leq a \vee x \geq d \end{cases} \tag{1.2}$$

The shape of this function is shown in Fig. 1.6b.

S-Shape MF

An S-shape MF is defined by its parameters $[a, b]$ such

$$\mu_A(x) = \begin{cases} \frac{x-a}{b-a} & a \leq x \leq b \\ 1 & b \leq x \\ 0 & x \leq a \end{cases} \tag{1.3}$$

The shape of this function is shown in Fig. 1.6c.

Z-Shape MF

A Z-shape MF is defined by its parameters $[c, d]$ such

$$\mu_A(x) = \begin{cases} 1 & x \leq c \\ \frac{d-x}{d-c} & c \leq x \leq d \\ 0 & x \geq d \end{cases} \tag{1.4}$$

The shape of this function is shown in Fig. 1.6d.

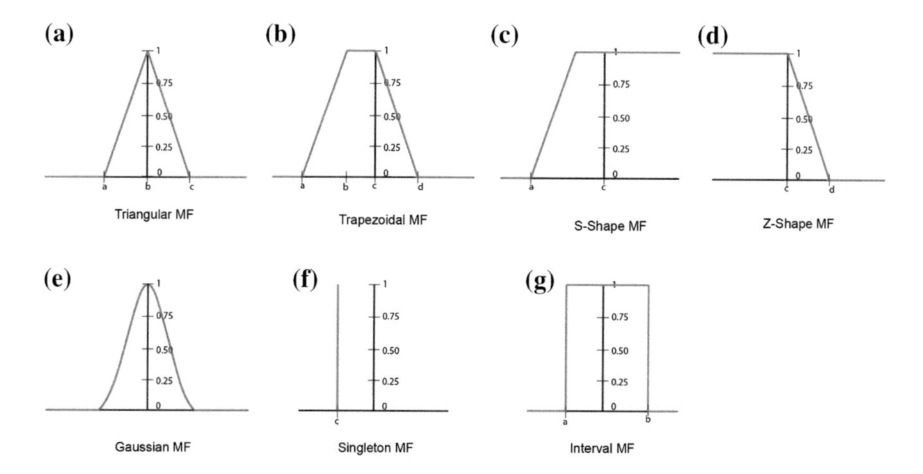

Fig. 1.6 Membership functions shapes

Gaussian MF
A Gaussian MF is defined by its parameters $[c, \sigma]$ such

$$\mu_A(x) = e^{-\frac{1}{2}\left(\frac{x-c}{\sigma}\right)^2} \tag{1.5}$$

The shape of this function is shown in Fig. 1.6e.

Singleton MF
A singleton MF is defined by its parameter c such as

$$\mu_A(x) = \begin{cases} 1 & x = c \\ 0 & \text{otherwise} \end{cases} \tag{1.6}$$

The shape of this function is shown in Fig. 1.6f.

Interval or Crisp MF
An interval or crisp MF is characterized by an interval [a, b]

$$\mu_A(x) = \begin{cases} 1 & a \le x \le b \\ 0 & \text{otherwise} \end{cases} \tag{1.7}$$

The shape of this function is shown in Fig. 1.6g; this membership shape is a particular case of T1FS where the membership is a crisp value as the traditional Boolean logic.

Such MF's are widely used in Type-1 FS for the fuzzification process (mapping from fuzzy to crisp value).

There are more possibly ways to assign membership values or functions to fuzzy variables than there are to assign probability density functions to random variables. This assignment process can be intuitive or it can be based on some algorithmic or logical operations. The following is a short list of methods described in the literature to assign membership values or functions to fuzzy variables.

1. Intuition—This method is simply derived from the capacity of humans to develop membership functions through their own innate intelligence and understanding.
2. Inference—It uses knowledge to perform deductive reasoning. That is to deduce or infer a conclusion, given a body of facts and knowledge.
3. Rank ordering—Assessing preferences by a single individual, a committee, a poll, and other opinion methods can be used to assign membership values to a fuzzy variable.
4. Inductive reasoning—An automatic generation of membership functions can also be accommodated by using the essential characteristic of *inductive reasoning,* which derives a general consensus from the particular (derives the generic from the specific). The induction is performed by the entropy minimization principle, which clusters most optimally the parameters corresponding to the output classes.

On the other hand, the operations on fuzzy sets are defined by means of their membership functions so that the definitions of those operations generalize their equivalents in crisp set theory.

Intersection: given the fuzzy sets A and B, defined in U with μ_A and μ_B membership functions and their intersection is the fuzzy set:

$$A \cap B = \{x, (\mu_{A \cap B}(x)) | x \in U\}, \quad \text{with } \mu_{A \cap B}(x) = \mu_A(x) \sim \wedge \mu_B(x)$$

Intersection operators belong to the class of Triangular Norms or T-Norms. These are binary operators defined in the interval [0, 1] satisfying the following properties:

$$T(0,1) = T(1,0) = T(0,0) = 0, T(1,1) = 1; \text{(agreement with boolean AND)}.$$
$$T(x,1) = x, \forall x \in [0,1]; \text{(identity)}.$$
$$T(x,y) = T(y,x), \forall x,y \in [0,1]; \text{(commutativity)}.$$
$$T(x1,y) \leq T(x2,y), \forall x1, x2, y \in [0,1], x1 \leq x2; \text{(monotocity)}.$$

Some operators have been proposed like T-Norms. The most used ones are the following:

$$\begin{aligned} \text{min}: \quad & \mu_{A \cap B}(x) = \mu_A(x) \sim \wedge \mu_B(x) = \min(\mu_A(x), \mu_B(x)); \\ \text{algebraic product}: \quad & \mu_{A \cap B}(x) = \mu_A(x) \sim \wedge \mu_B(x) = \mu_A(x)\mu_B(x); \\ \text{bounded difference}: \quad & \mu_{A \cap B}(x) = \mu_A(x) \sim \wedge \mu_B(x) = \max(0, \mu_A(x) + \mu_B(x) - 1); \end{aligned}$$

Union: given the fuzzy sets A and B, defined in U with μ_A and μ_B membership functions and their union is the fuzzy set:

$$A \cup B = \{x, (\mu_{A \cup B}(x)) | x \in U\}, \quad \text{and } \mu_{A \cup B}(x) = \mu_A(x) \sim \vee \mu_B(x),$$

Union operators belong to the class of S-Norms or Triangular Co-Norms (T-Co-Norms). These are binary operators defined in the interval [0, 1] satisfying the following properties:

$$S(0,1) = S(1,0) = S(1,1) = 1, S(0,0) = 0; \text{(agreement with boolean OR)}.$$
$$S(x,0) = x, \forall x \in [0,1]; \text{(identity)}.$$
$$S(x,y) = S(y,x), \forall x,y \in [0,1]; \text{(commutativity)}.$$
$$S(x1,y) \leq S(x2,y), \forall x1, x2, y \in [0,1], x1 \leq x2; \text{(monotonicity)}.$$

Some operators have been proposed as T-CoNorms. The most used are following:

$$\begin{aligned} \text{max}: \quad & \mu_{A\cup B}(x) = \mu_A(x) \sim \vee \mu_B(x) = \max(\mu_A(x), \mu_B(x)); \\ \text{algebraic sum}: \quad & \mu_{A\cup B}(x) = \mu_A(x) \sim \vee \mu_B(x) = \mu_A(x) + \mu_B(x) - \mu_A(x)\mu_B(x); \\ \text{bounded sum}: \quad & \mu_{A\cup B}(x) = \mu_A(x) \sim \vee \mu_B(x) = \min(1, \mu_A(x) + \mu_B(x)); \end{aligned}$$

Complement: given the fuzzy set A defined in the Universe of Discourse U, its complement is the fuzzy set:

$$\overline{A} = \left\{x, (\mu_{\overline{A}}(x))|x \in U\right\} \quad \text{and} \quad \mu_{\overline{A}}(x) = \sim\neg(\mu_A(x)) = 1 - \mu_A(x)$$

where $\sim\neg$ denotes the "fuzzy negation" of a membership function, coincident in this case with the complement operator for crisp sets. There are others complement operators $\sim\neg$. In general, they are C-Norms.

1.3.3 Discourse Universe and Membership Degree

If A is a Type-1 FS and each set has a set of membership values $\mu_A(x_i)$ for each x_i, all the values $x_i \in X$ are called the discourse universe, or simply the discourse. The discourse universe can be a set of ordered or nonordered values.

Each discourse value is mapped by f and f is a function, then each value characterized by f is a membership degree, which is related to the belonging to a certain set. The membership degree has values in interval (0, 1), where a crisp value can belong to a set in 100 % if its membership degree is equal to 1 [54].

1.4 Extension Principle

The extension principle is a basic concept of fuzzy sets theory used for generalizing the crisp mathematical concepts into fuzzy sets. It was implied in the first fuzzy logic paper [40] and after that some modifications was developed [51, 55] until reach the present concept.

Consider two universes of discourse X and Y, and a mapping function of the form $y = f(x)$. Consider also a collection of elements from the set A which are defined in the universe x. The extension principle defines the image of fuzzy set A on Y under the mapping f. If the image is denoted as B. The mapping is $B = f(A)$.

The membership functions which defines B on the universe of a membership interval [0, 1] is:

$$\mu_B(y) = \bigvee_{f(x)=y} \mu_A(x)$$

The fuzzy vectors are defined as vector containing membership values that utilize matrix relations. The fuzzy sets A and B are expressed in vector forms as follows:

$$a = \{a_1, \ldots, a_n\} = \{\mu_A(x_1), \ldots, \mu_A(x_n)\} \quad \text{for} \quad i = 1, 2, \ldots, n$$

and

$$b = \{b_1, \ldots, b_n\} = \{\mu_B(y_1), \ldots, \mu_B(y_m)\} \quad \text{for} \quad i = 1, 2, \ldots, m$$

Now the image of the fuzzy set A can be determined through the use of composition operators $B = A \circ R$, where R is a $n \times m$ fuzzy relation matrix. The mapping can be defined then as $B = f(A_1, A_2, \ldots, A_n)$ where the membership function of the image B is given by:

$$\mu_B(y) = \max_{y=f(x_1,x_2,\ldots,x_n)} \{\min[\mu_{A_1}(x_1), \mu_{A_2}(x_2), \ldots, \mu_{A_n}(x_n)]\}$$

The previous equation is called the Zadeh's extension principle and is expressed for a discrete-valued function f. If the functions are continuous, the max operator is replaced by the sup (supremum or last upper bound) operator.

Table 1.2 Basic identities using the basic fuzzy logic operations

Name	Symbol
Contradiction	$A \sqcap \neg A = \emptyset$
Excluded middle	$A \sqcup \neg A = X$
Idempotency	$A \sqcap A = A, A \sqcup A = A$
Involution	$\neg(\neg A) = A$
Commutativity	$A \sqcap B = B \sqcap A, A \sqcup B = B \sqcup A$
Associativity	$(A \sqcup B) \sqcup C = A \sqcup (B \sqcup C)$, $(A \sqcap B) \sqcap C = A \sqcap (B \sqcap C)$
Distributivity	$A \sqcup (B \sqcap C) = (A \sqcup B) \sqcap (A \sqcup C), A \sqcap (B \sqcup C) = (A \sqcap B) \sqcup (A \sqcap C)$
Absorption	$A \sqcup (A \sqcap B) = A, A \sqcap (A \sqcup B) = A$
Complement absorption	$A \sqcup (\neg A \sqcap B) = A \sqcup B, A \sqcap (\neg A \sqcup B) = A \sqcap B$
DeMorgan law	$\neg(A \sqcup B) = \neg A \sqcap \neg B, \neg(A \sqcap B) = \neg A \sqcup \neg B$

1.4.1 Basic Identities

Using the basic fuzzy properties [53] and [54], some identities are inherited from the Boolean logic shown in Table 1.2.

1.5 Fuzzy Logic Rules

An important concept in fuzzy logic is a *fuzzy proposition.* Fuzzy propositions represent statements like "V is big," where "big" is a linguistic label, defined by a fuzzy set on the universe of discourse of variable V. *Fuzzy (linguistic) labels* are also referred to as *fuzzy constants, fuzzy terms or fuzzy notions*. Fuzzy propositions connect variables with linguistic labels defined for those variables. Fuzzy relations can also be assembled from linguistic knowledge, expressed as if—then rules. Such knowledge may come from experts, from polls, or from consensus building. Relations also arise from notions of classification where issues associated with similarity are central to determining relationships among patterns or clusters of data. By using the basic properties and operations defined for fuzzy sets, any compound rule structure may be decomposed and reduced to a number of simple canonical rules. These rules are based on natural language representations and models, which are themselves based on fuzzy sets and fuzzy logic. The fuzzy level of understanding and describing a complex system is expressed in the form of a set of restrictions on the output based on certain conditions of the input. Restrictions are generally modeled by fuzzy sets and relations. These restriction statements are usually connected by linguistic connectives such as "and," "or," or "else." The restriction R1, R2, …, Rr apply to the output actions, or consequents of the rules. The fuzzy rules are used for stabilizing a relationship between the inputs and the outputs, in 1973, Zadeh published [55], in this paper was presented a new approach for analyzing complex systems based on capturing the human knowledge in the form of fuzzy rules.

A fuzzy rule can be defined as a conditional statement in the form:

IF x IS A
THEN y IS B

where x is an input variable and A is a defined fuzzy set defined in the discourse X; on the other hand, y is an output variable and B is its corresponding fuzzy set defined in the discourse Y.

If the output depends on more than one variable, there must be a relationship between the inputs; for example, consider a brake system defined from a fuzzy database with the following rule:

IF *speed* is **high** AND *breaks_wear* is **low**
THEN *braking_force* is **medium**

When an operator between the inputs is used, all the parts of the antecedents are calculated simultaneously.

Another important operator for fuzzy rules is the OR; consider the next example:

IF *work* is **Low** OR *time_operating* a machine is **medium**
THEN *maintenance_schedule* is **regular**

There is also the possibility of having certain situations where more than one output is activated as:

IF *temperature* is **High**
THEN:
hot_water is **reduced** ***and*** *cold_water* is **increased**

1.6 Defuzzification Methods

This is the last step in a fuzzy logic inference; for instance, if the actuators of the system are activated, a crisp value of voltage has to be activated. The defuzzification aggregates the outputs into a single crisp number; there are several methods for doing this and the most commons are the shown below:

Centroid
This is the most widely used defuzzification method in T1FLS, also known as the Centroid of Area (COA). For this defuzzification method, the shape of the T1 Inferred Set $\mu_B(y)$ is needed. Hence, the expression can be expressed by:

$$y = \frac{\sum_{i=1}^{N} y * \mu_B(y)}{\sum_{i=1}^{N} \mu_B(y)} \tag{1.8}$$

Bisector
This defuzzification method searches for y satisfying the following condition:

$$\sum_{i=\alpha}^{y'} \mu_B(y) = \sum_{i=y'}^{\beta} \mu_B(y) \tag{1.9}$$

where $\alpha = min(Y|\mu_B(y) \neq 0), \beta = \max(Y|\mu_B(y) \neq 0)$, such $\alpha, \beta \in Y$. This means that a vertical line $y = y'$ partitions the region between $[\alpha, \beta]$ and $[0, \mu_B(y)]$ into two regions with the same area.

Mean Of Maximum
Mean of maximum considers the crisp output as the mean of each

$$Y' = \left[\mathrm{y_i}, \mathrm{y_{i+k}}\right]$$

such as:

$$\begin{aligned} \mu_\mathrm{B}(\mathrm{y_i}) = \mu_\mathrm{B}\left(\mathrm{y_{i+k}}\right) = \max(\mu_\mathrm{B}) : \\ \mathrm{y} = \frac{1}{2}\left(\mathrm{y_i} + \mathrm{y_{i+k}}\right) \end{aligned} \tag{1.10}$$

It must be searched for the first $\mathrm{i}|\mathrm{y_i}\in\mathrm{Y}$, such $\mu_\mathrm{B}(\mathrm{y_i}) = \max(\mu_\mathrm{B})$ starting from left to right, i.e. from 1 to N. Also, you must search for the last $\mathrm{i} = \mathrm{i} + \mathrm{k}, \mathrm{k}|\mathrm{y_{i+k}}\in\mathrm{Y}$, such $\mu_\mathrm{B}\left(\mathrm{y_{i+k}}\right) = \max(\mu_\mathrm{B})$ starting from right to left, i.e. from N to 1.

Height
The height defuzzifier, also known as the Center Average defuzzifier, searches for $y = y^l \in Y$, such $\mu_B\left(y^l\right)$ is the maximum for each consequent set, i.e., before every trimmed set is aggregated in the inferred set search. If several values $Y' = \left[y_j, y_k\right] \in Y$, present the maximum membership value for that consequent set, the average between its ends is selected, i.e.

$$y^l = \frac{1}{2}\left(y_j + y_k\right) \tag{1.11}$$

Finally, whenever each y^l and its corresponding $\mu_{B^l}\left(y^l\right)$ is already known, then the centroid is calculated as:

$$y = \frac{\sum_{l=1}^{M} y^l * \mu_{B^l}\left(y^l\right)}{\sum_{l=1}^{M} \mu_{B^l}\left(y^l\right)} \tag{1.12}$$

Center Of Sums
This method combines the consequent MF area and its centroid.

For discrete discourse universes Y, such $\Delta y = y_i - y_{i-1}\in Y$, the proposed area can be computed as the sum of each consequent membership degree, i.e.:

$$a_{B^l} = \sum_{i=1}^{N} \mu_{B^l}(y_i) \tag{1.13}$$

Then the center of sum can be computed as

$$y_a = \frac{\sum_{l=1}^{M} c_{B^l} * a_{B^l}}{\sum_{l=1}^{M} a_{B^l}} \tag{1.14}$$

where $\mathrm{c_{B^l}}$ is the centroid of each consequence.

1.7 Fuzzy Inference Methods

Fuzzy inference is defined as the process of mapping an output from a given input using membership degrees; up to now, there were defined the steps of the inference models. So, it is required to define a block diagram with all the steps involved in the fuzzy logic inference. The steps expressed in a block diagram are shown in Fig. 1.7.

The most common fuzzy inference methods are: the Mamdani and Sugeno type that are shown in Fig. 1.8.

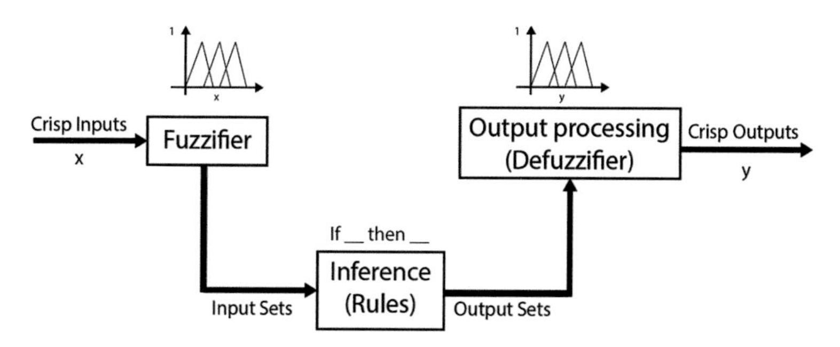

Fig. 1.7 Algorithm structure of Type-1 and Type-2 FLS

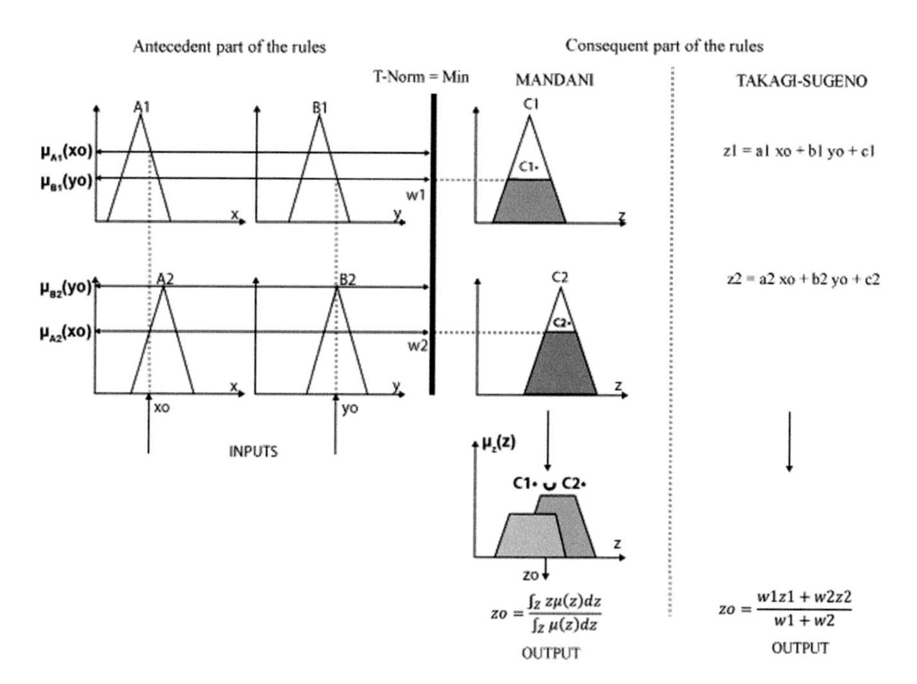

Fig. 1.8 Mamdani and Sugeno inference

Mamdani Inference

Ebrahim Mamdani [44] proposed a controller based on linguistic rules and fuzzy sets. This is one of the most used inferences methods in fuzzy logic. A rule in the Type-1 Mamdani Inference Model is expressed as:

$$\text{if}\, x_1 \, is \, A_1^1 \quad \text{and} \quad x_2 \,\text{is}\, A_2^1 \,\text{then}\, y \,\text{is}\, B \tag{1.15}$$

where the subindexes in A_1^1 refer to the number of the input variable and the super-indexes refer to the number of the set of the input variable labeled in the sub-index.

Its inference model could be based on the min–max or max-prod composition according to the rule set. Generally, the inference model can be performed with any composition of fuzzy *T-Norms and T-Conorms*. Generally, the fuzzy composition can be expressed by:

$$\mu_{\mathcal{R}_1 \circ \mathcal{R}_2}(x, z) = S\big(T\left(\mu_{\mathcal{R}_1}(x, y), \mu_{\mathcal{R}_2}(y, z)\right)\big) \tag{1.16}$$

where S is applied to the result of all the T-Norms in the relations. The compositions are constructed by using conjunctions and disjunctions operators, the compositions share the properties. From, the expert may create an inference stage of: a Single Rule with a Single Premise, a single rule with multiple premises and, a multiple rule with multiple premises [44]. In general, the resulting Type-1 FS obtained after the T-Norm part can be called as the firing strengths or the fulfillment degree from each rule.

Type-1 FS is called the Type-1 Inferred Set. Some defuzzification methods can be used to convert the Type-1 FS into a crisp output, which is the system output. The previous defuzzification methods are widely used in several applications. Some of them are more complex to the others. Now, suppose that the output discourse universe Y is discrete, Y is the crisp output, such $\mathrm{y} \in \mathrm{Y}$ where the Type-1 Inferred Set $\mu_{\mathrm{B}}(\mathrm{y})$ resides.

The first inference method, due to Mamdani and Assilian (1975), is the most common in practice and in the literature.

For clarifying the steps in the Mamdani method, an example is shown below:

Step 1: **Fuzzification**

It transforms the physical input values (position, voltage, degree, and so on) into a normalized fuzzy subset consisting of a subset (interval) for the range of the input values and a normalized membership function describing the degree of membership of the input belonging to this range. In the example, the first step is to transform the crisp inputs taken from the sensors, estimators, or measurements x_1, y_1 and limited within the universe of discourse X and Y, respectively. The crisp inputs are mapped to fuzzy values by the membership functions. In Fig. 1.8, a system with two inputs and one output is shown; the fuzzification consists in finding the membership values for each input variable. As shown in the example in Fig. 1.9, the system has

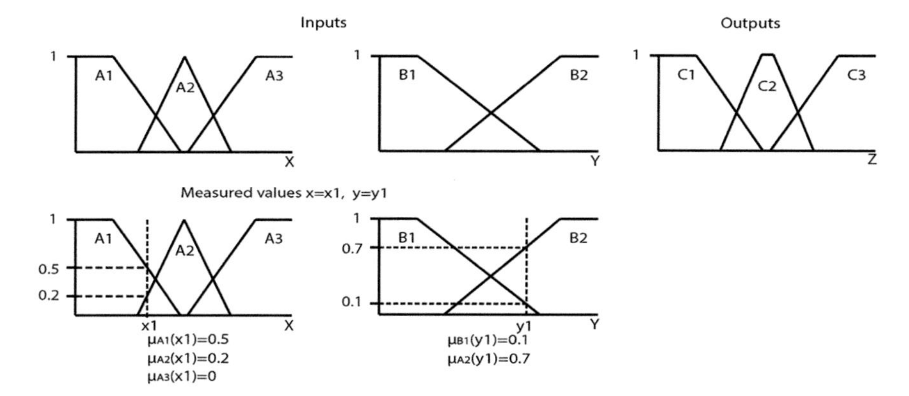

Fig. 1.9 Fuzzification step

two inputs x1 and y1 (it can be observed the membership values found by the membership functions for A1 = 0.5, A2 = 0.2 and A3 = 0 and for Y1 = 0.1 and Y2 = 0.7). Fuzzification implies to select reasonable and good, ideally optimal, and membership functions under certain convenient criteria meaningful to the developed application. Hence, a correct selection of shape and number of membership functions has to be done in this step.

Step 2: **Rules evaluation**

Designing a good fuzzy logic rule base is key to obtain a satisfactory response for a particular application. In the example, the rules are evaluated according to the measurements inputs; in this example, the AND operator is governed by the MIN T-norm, so when a rule has the AND operator, the value taken for evaluation of the output is the minimum membership value of the inputs. In the case of the OR operator, the MAX T-conorm is implemented and therefore the taken value for evaluation of the outputs is the maximum membership value of the implicated inputs. The third rule does not relate the inputs and hence the membership value of the inputs is taken for the evaluation of the output. The mim–max inference is applied in the example. In the inference method knowledge was used to perform deductive reasoning. That is, it is based on deduce or infer a conclusion, given a body of facts and knowledge (rules). The min–max inference approach is a technique that can be used to evaluate the relationship between premise and antecedents in a Fuzzy logic system. Figure 1.10 shows the rule evaluation step. The causal relations between concepts are often defined by linguistic variables, which are words that describe the strength of the relationship. The min–max inference approach can be utilized to evaluate these linguistic variables. The minimum value of the links in a path is considered to be the path strength. If more than one path exists between the cause variable and the effect variable, the maximum value of all the paths is considered to be the overall effect. In other words, the indirect effect amounts to specifying the weakest linguistic variable in a path, and the total effect amounts to specifying the strongest of the weakest paths.

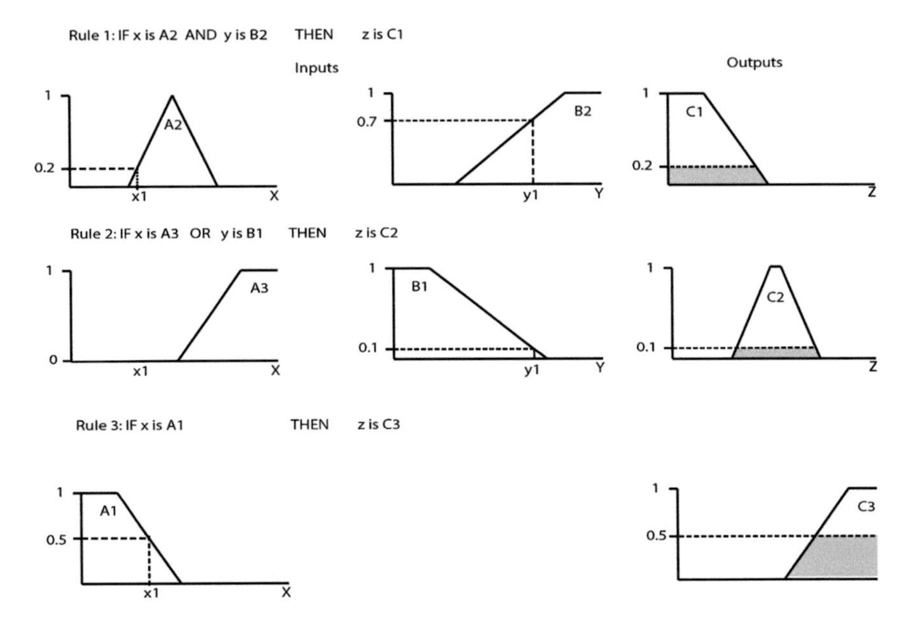

Fig. 1.10 Rule evaluation step

Step 3: **Aggregation of the rules outputs and defuzzification**

The defuzzification module is in a sense the reverse of the fuzzification module: it converts all the fuzzy terms created by the rule base of the fuzzy logic system to crisp values and then this value can be sent to a physical system (plant, process), so as to execute the fuzzy logic controller. It creates a crisp, overall output signal, by combining all possible outputs from the rule base into a weighted average formula, such as the center of sum. The defuzzification step is illustrated by Fig. 1.11.

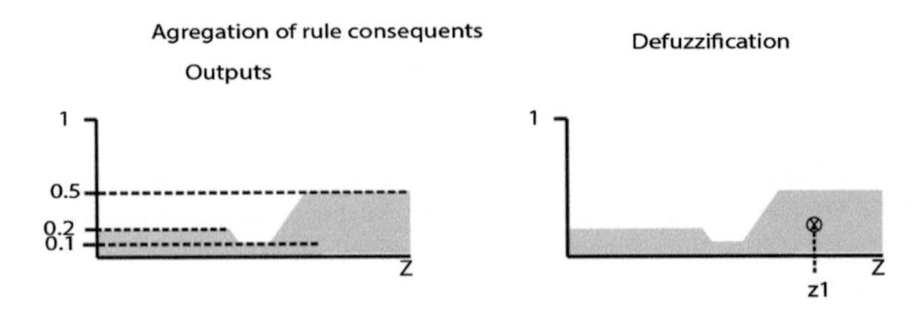

Fig. 1.11 Defuzzification step for Mamdani inference model

1.8 Takagi-Sugeno-Kang

The Sugeno inference is different from Mamdani because the defuzziciation uses a polynomial equation. So, the Sugeno if–then rules are described in a different form: the consequents of those rules could be (linear) functions. Takagi Sugeno and Kang in 1985 [48], proposed a mathematical model for inference systems based on polynomials, called the Takagi-Sugeno-Kang Inference Model (TSK). Their purpose was to develop a systematic approach to generate fuzzy rules from a given input-output dataset. Typically, a T1 TSK rule seems alike:

$$if\ x_1\ is\ A_1^1\ AND\ x_2\ is\ A_2^1\ THEN\ y = f(x_1, x_2) \tag{1.17}$$

Figure 1.12 shows a complete representation of Sugeno inference in this representation; it is possible to observe the defuzzification based on polynomials.

where f is a crisp function defined by polynomials of input variable values x_1 and x_2, but generally it can be any appropriated function which can satisfy the required outputs within the fuzzy region specified by the rule premise. Also, each rule may have a different polynomial. The polynomial can be of any order, but usually the polynomial can be of second, first, and zero order. The zero-order TSK model can be considered as a particularity, because this behaves like a Mamdani Inference model using singletons as consequent sets.

For Type-1 FL, the firing strengths are related to their corresponding polynomials. The consequent FS y^i are polynomials, where $I = 1, 2, \ldots, M$ and $j = 0, 1, \ldots, p$. M is the total number of firing strengths and p is the total number of crisp input variables.

$$F^i = T\left(\mu_{f_1}(x_1), \mu_{f_1}(x_1), \ldots \mu_{f_p}(x_p)\right) \tag{1.18}$$

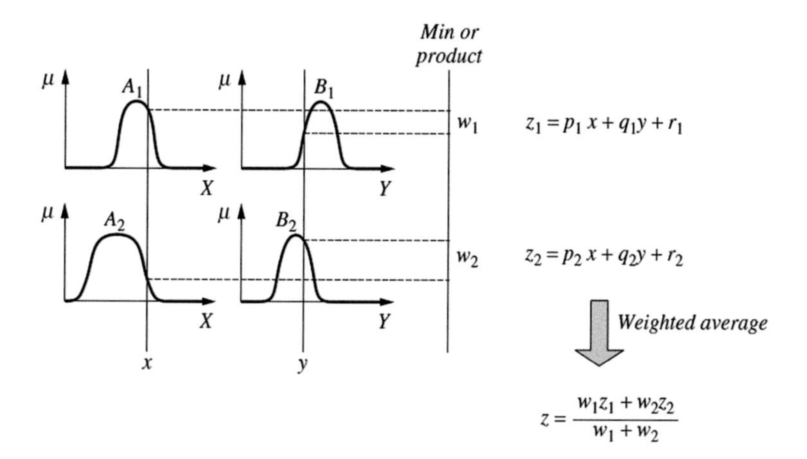

Fig. 1.12 Sugeno inference method

Each firing strength can be computed as

$$y^i = \sum_{k=1}^{p} \left(a_k^i x_k^2 + b_k^i x_k + c_k^i \right) \tag{1.19}$$

So, the TSK output is:

$$y_{\mathrm{TSK}} = \sum_{y^1 \in Y} \cdots \sum_{y^M \in Y} \sum_{f^1} \cdots \sum_{f^M} \frac{\sum_{i=1}^{M} f^i y^i}{\sum_{i=1}^{M} f^i} \tag{1.20}$$

For the purpose of clarifying the TSK inference method an example is shown below using constant values in the membership functions (singleton functions)

Step 1: **Fuzzifier**.
As in the Mamdani type, the first step is to evaluate the measured values and obtain the crisp value. Notice that the definition of the outputs is only a singleton MF, by doing this the computation of the centroid of mass or any other defuzzification method is avoided because there is only one value defined for the output (see Fig. 1.13).

Step 2: **Rules evaluation**.
The evaluation of the rules is presented in Fig. 1.14. In this rule evaluation, an inference method is applied (e.g., min-max).

Step 3: **Aggregation of the rules outputs anddefuzzification**.
The next step consists of the evaluation of the output (see Fig. 1.15). If $Z1 = 1$, $Z2 = 2$, and $Z3 = 3$.

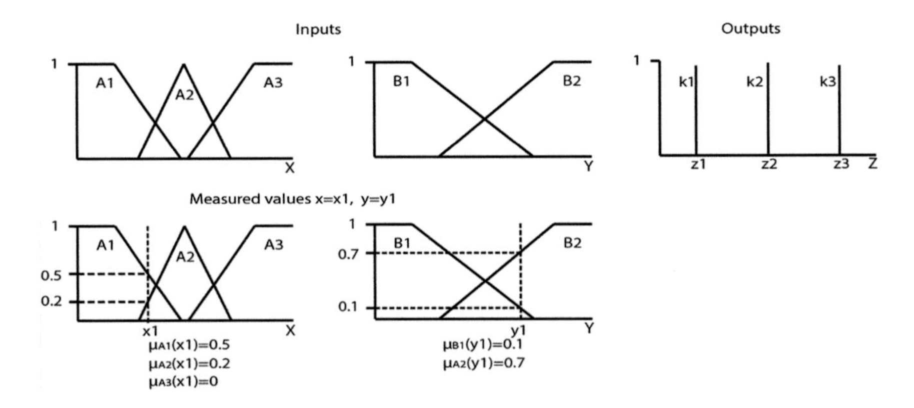

Fig. 1.13 Fuzzification step

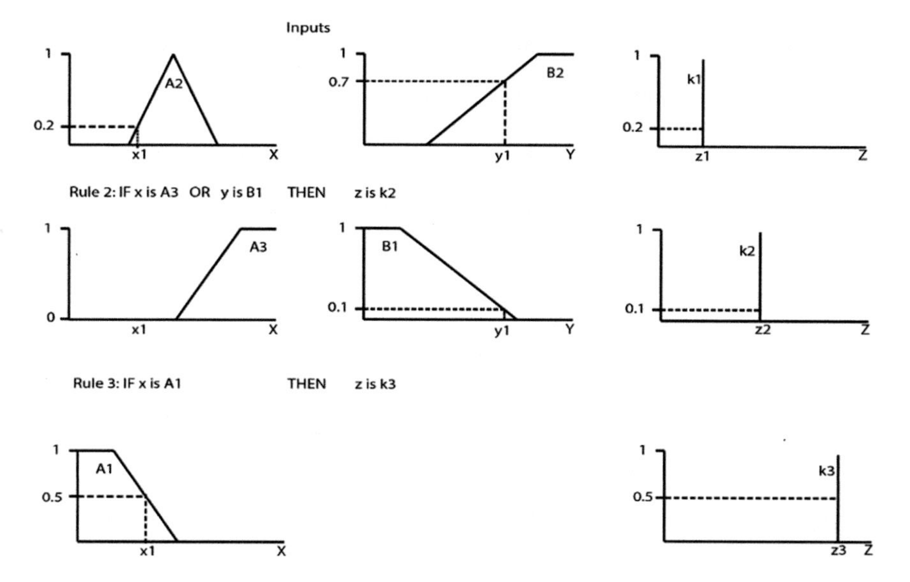

Fig. 1.14 Rule evaluation step

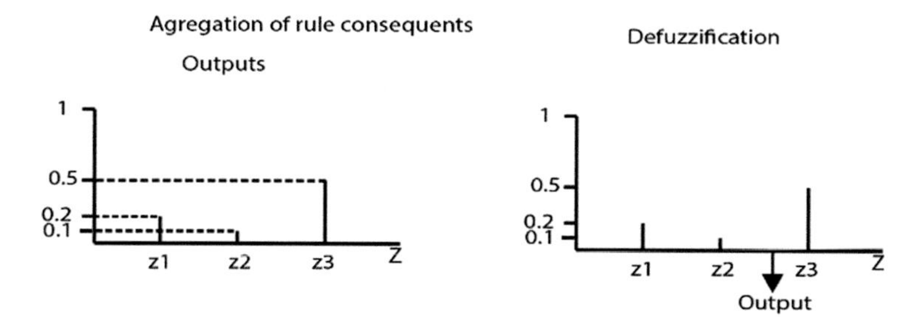

Fig. 1.15 Defuzzification step for TSK inference model

The output is calculated by

$$\text{Output} = \frac{(0.2 * 1) + (0.1 * 2) + (0.5 * 3)}{(0.5 + 0.2 + 0.1)}$$

An example about fuzzy logic control is presented below.

IF TEMPERATUREIS HIGH ANDPRESSURE IS LOWTHEN SET THEVALVE TO K_1

The heuristic rule for temperature and pressure control is derived from people experience. Both, temperature and pressure, and valve setting are the fuzzy variables that focus on the corresponding fuzzy sets. Low, high, and medium settings

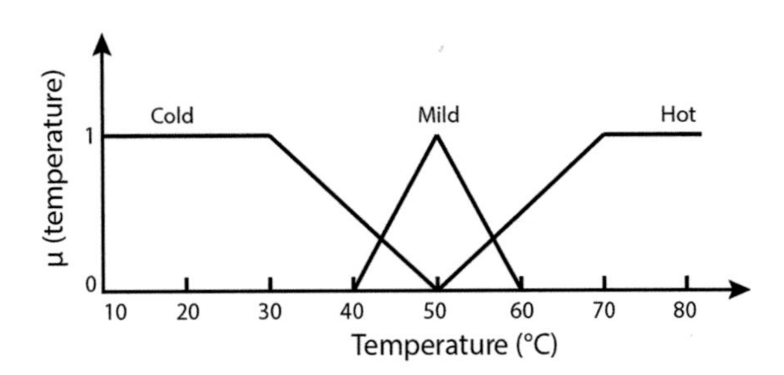

Fig. 1.16 Fuzzy sets of temperature

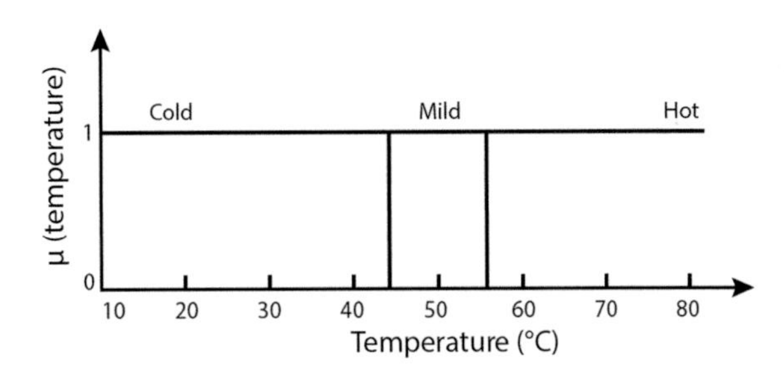

Fig. 1.17 Crispy sets of temperature

have the characteristic membership functions and also are fuzzy values determined by the user. Figure 1.16 shows the fuzzy sets of the example, as well as, Fig. 1.17 shows up the crispy sets. Normally, it is easier to adjust the membership functions and the polynomials in the outputs than membership functions in the input and outputs, so Sugeno is proposed to be represented by adaptive networks (ANFIS) that help to design a fuzzy logic controller.

A system rule could be made by:

$$\text{IF TEMPERATURE} < 60\,^{\circ}\text{C} > 25\,\text{PSI}$$
$$\text{THEN SET THE FUE VALVE } 40\,\%$$

where pressure, temperature, and fuel valve settings are the parameters. The rules processes the logic variables generated by the input conditions. Fuzzy controls require fewer rules compared to expert system. Systems rules depend on the parameters established by the user, such like the amount of them.

1.9 Numerical Example (Mandani)

Consider a humidity fuzzy controller of a microbiological incubator. The inputs of the system are the relative humidity of the air and the temperature. The fuzzy sets of the input can be seen in the Fig. 1.18.

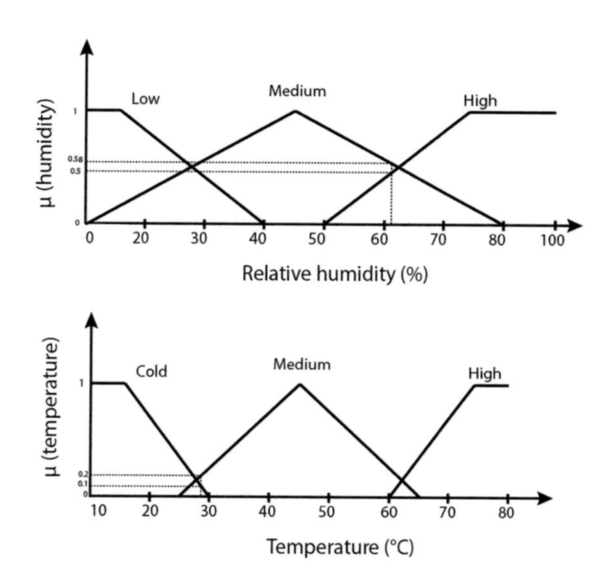

Fig. 1.18 Relative humidity and temperature variables

The measures taken from the sensors are 61 % for the relative humidity and 29 °C. Thus, the membership value of the medium set for humidity is cut on 0.58, and the high set is cut on 0.5. For the temperature, the cold set is cut on 0.2 and the medium set cut on 0.1. Figure 1.19 shows the membership functions of the output.

$$\mu(r\_\text{humidity} = \text{low}) = 0$$
$$\mu(r\_\text{humidity} = \text{medium}) = 0.58$$
$$\mu(r\_\text{humidity} = \text{high}) = 0.5$$
$$\mu(\text{temperature} = \text{cold}) = 0.1$$
$$\mu(\text{temperature} = \text{medium}) = 0.2$$
$$\mu(\text{temperature} = \text{high}) = 0$$

Consider that the following rules are activated with the activated fuzzy sets of the inputs:

IF TEMPERATURE IS **COLD** AND RELATIVE HUMIDITY IS **MEDIUM** THEN SPEED IS **LOW**

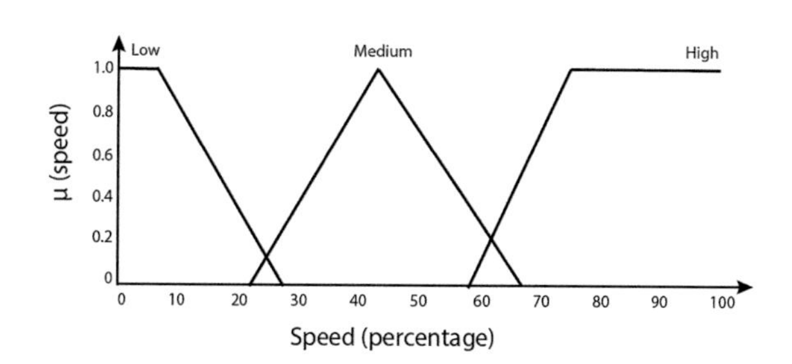

Fig. 1.19 Speed variable

$$\mu_{\text{speed}}(\text{low}) = \min[\mu(\text{temperature} = \text{cold}), \mu(r\_\text{humidity} = \text{medium})] \\ = \min[0.1, 0.58] = 0.1$$

IF TEMPERATURE IS **MEDIUM** AND RELATIVE HUMIDITY IS **MEDIUM** THEN THE SPEED IS **MEDIUM**

$$\mu_{\text{speed}}(\text{medium}) = \min[\mu(\text{temperature} = \text{medium}), \mu(r\_\text{humidity} = \text{medium})] \\ = \min[0.2, 0.58] = 0.2$$

IF TEMPERATURE IS **HIGH** AND RELATIVE HUMIDITY IS **HIGH** THEN THE SPEED IS **HIGH**

$$\mu_{\text{speed}}(\text{high}) = \min([\mu(\text{temperature} = \text{high}), \mu(r_{\text{humidity}} = \text{high})] \\ = \min[0, 0.5] = 0$$

The outputs are then cut in the membership value (dashed lines) in Fig. 1.20. And the center of mass can be calculated as it is shown in Fig. 1.21.

The center of mass can be obtained with the Eq. 1.17, then the crisp value of the output is define.

$$y = \frac{(0+10+20)*0.1+(30+40+50+60)*0.2+(70+80+90+100)*0}{0.1+0.1+0.1+0.2+0.2+0.2+0.2+0+0+0+0} \\ = 34.45$$

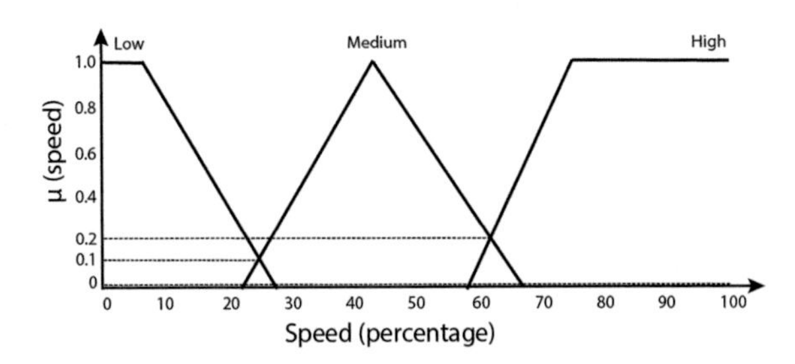

Fig. 1.20 Speed variable

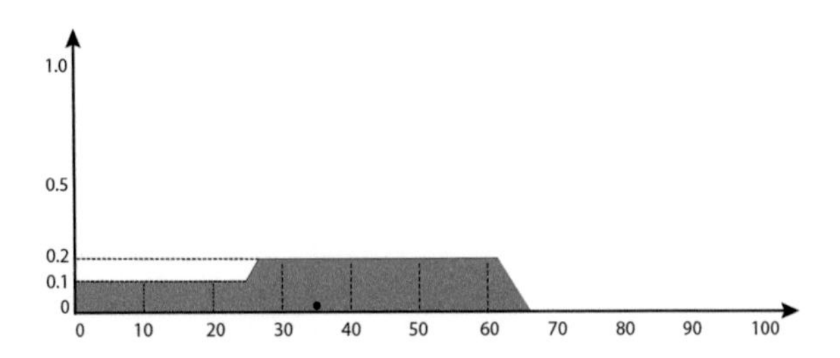

Fig. 1.21 Speed variable

1.10 Basic Numerical Example (TSK)

Consider that the previous controller uses a TSK inference model for simplifying the computation of the output. The same input shapes are used and the rules are the same. The next figure shows the output points selected:

The following rules are activated and the results are:

IF TEMPERATURE IS **COLD** AND RELATIVE HUMIDITY IS **MEDIUM** THEN SPEED IS **LOW**

$$\begin{aligned}\mu_{\text{speed}}(\text{low}) &= \min[\mu(\text{temperature} = \text{cold}), \mu(r\_\text{humidity} = \text{medium})] \\ &= \min[0.1, 0.58] = 0.1\end{aligned}$$

IF TEMPERATURE IS **MEDIUM** AND RELATIVE HUMIDITY IS **MEDIUM** THEN THE SPEED IS **MEDIUM**

$$\begin{aligned}\mu_{\text{speed}}(\text{medium}) &= \min[\mu(\text{temperature} = \text{medium}), \mu(r\_\text{humidity} = \text{medium})] \\ &= \min[0.2, 0.58] = 0.2\end{aligned}$$

IF TEMPERATURE IS **HIGH** AND RELATIVE HUMIDITY IS **HIGH** THEN THE SPEED IS **HIGH**

$$\begin{aligned}\mu_{\text{speed}}(\text{high}) &= \min([\mu(temperature = \text{high}), \mu\left(r_{\text{humidity}} = \text{high}\right)] \\ &= \min[0, 0.5] = 0\end{aligned}$$

The outputs are then cut in the membership value (dashed lines) in Fig. 1.22. And the crisp output value is computed as follows:

$$y = \frac{5 * 0.1 + 45 * 0.2 + 90 * 0}{0.1 + 0.2 + 0} = 31.67$$

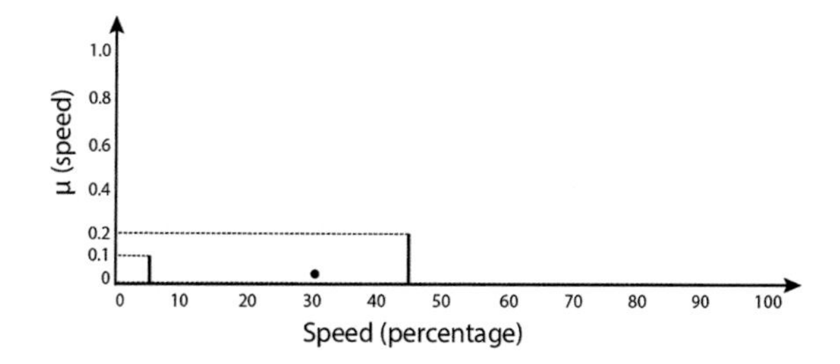

Fig. 1.22 Speed variable

1.11 Type-2 Fuzzy Logic Set

1.11.1 Historical Review of Advances

Fuzzy logic sets [40] were developed by Zadeh and published in 1965 with the objective of dealing with the uncertainties and imprecision intrinsic in a complex system where crisp outputs of the traditional logic operators do not express the desired information that defines the system; instead, he thought that the systems could be represented with certain degree of membership and established the mathematical bases for logic operators on a fuzzy environment. His work has been extensively tasted in fields like control theory and artificial intelligence applications and its functionality has been proved. However, when more challenging problems take place like imprecise or noisy input or statistical uncertainties from different expert's opinions on the expertise database, the performance of T1 fuzzy logic sets is not suitable for implementation.

For solving this, Mizumoto and Tanaka [46] studied and presented the basic algebraic structures for fuzzy sets for the joint, meet, and negation operations employing the principles of fuzzy sets defined by Zadeh; after that Karnik and Mendel [56] developed the mathematical bases for the implementation of type 2 fuzzy logic sets. In [56] is defined the concept of footprint of uncertainty, the mathematical bases for the T-norm and T-conorm and three type reducer techniques for implementing the defuzzification: centroid, height, and center of sets.

In 1985 and 1988, Takagi Sugeno and Kang [48, 49] proposed a new approach for developing a systematic methodology for generating fuzzy rules from a given input–output database of T1FS. Later on, it was also applied for T2FS.

Once the bases of type 2 were published, Gorzalczany [57] published the first interval type 2 technique and many other authors started implementing it and continue researching [58–60]. In 2000, Liang and Mendel published the theory and design of the interval type 2 fuzzy sets [61] and introduced the concept of upper and lower membership functions; this paper presented an application for performing time series forecasting with the input signal corrupted by additive noise.

Interval type 2 fuzzy logic has developed quickly due to the fast computation respect to the general case where the secondary set has a noninterval shape making the computation much more difficult. The fast success of interval type 2 fuzzy logic is proved by the raise in the number of publications since 1999 and until present. Interval type 2 fuzzy logic provides good applications performance and there are many comparisons with type 1 fuzzy logic; however, the great potential of the general case gets limited when interval is used for defining the secondary membership function.

Despite the better performance in terms of computation time of the IT2FLS, for real-time applications it was not fast enough; so Wu and Mendel [62] worked on a new method for computing the type reduced set based on uncertain bound sets avoiding the calculation of the centroid for defuzzification. Mendel [63] continued working on the improvement of the computation time for IT2FLS with an algorithm based on the symmetrical geometry of the fuzzy sets doing half of the operations that was made before and therefore reducing the computation time by 50 %. In 2008, Coupland and John [64] published a method fast defuzzification based on geometric representations and operations which claims to be up to 200,000 times faster that type reduction in order to implement general T2FLS.

Wu and Tan [45] proposed a new simplification architecture technique for improving the performance of the traditional type reduction which is prohibit in real-time applications; later, Wu and Mendel [65] worked together for improving the original Karnik–Mendel algorithm in terms of number of iteration for convergence, the results show more than 39 % save in computation time and they called it Enhanced Karnik Mendel (EKM).

Lui [50] proposed a new type reduction method based on a new representation called α-plane representation in order to greatly reduce the computation complexity from exponentially to linear. The α-plane representation is the union of all primary membership functions whose secondary grades are equal or greater than alpha, which avoid unnecessary computation of the algorithm to converge to a real value. Recently, Wu et al. published another fast method for computing the centroid of a type 2 fuzzy set [66]. Which is faster than the EKM and Lui's method, the concept of alpha-plane representation is taken from the Lui's work but they discovered and applied other useful properties of the α-plane which improve the computation speed.

1.11.2 Type-2 Fuzzy Sets (T2FS)

T2FS are the natural extension of type 1 but they provide additional information in the secondary membership function, that is, they not only depend on a variable "x" but an additional variable "u", fuzzy sets are represented by $\tilde{A}$ [67] :

$$\tilde{A} = \left\{ \left((x, u), \mu_{\tilde{A}}(x, u) \right) | \, \forall x \in X \quad \forall u \in J_x \subseteq [0, 1] \right\} \tag{1.21}$$

where $\mu_{\tilde{A}}(x, u)$ is a type 2 membership function that has a value between $0 < \mu_{\tilde{A}}(x, u) < 1$.

In order to visualize the new dimension (see Fig. 1.23), for a general case T2FS, the degree of membership depends on the variables x and u; in order to fast visualization the T2FS are usually represented in two dimensions and the plane x-J is shown and the area which is seen is called the footprint of uncertainty. The additional dimension gives an extra degree of uncertainty and therefore some positions of the membership function have a higher degree of membership than others; this can result in an additional problem for computation because of the huge amount of operations required. For leading with this interval type 2 fuzzy sets (IT2FS) are the most commonly used shape of T2FS; these are special cases where the secondary membership degree is constant for all the universe of discourse

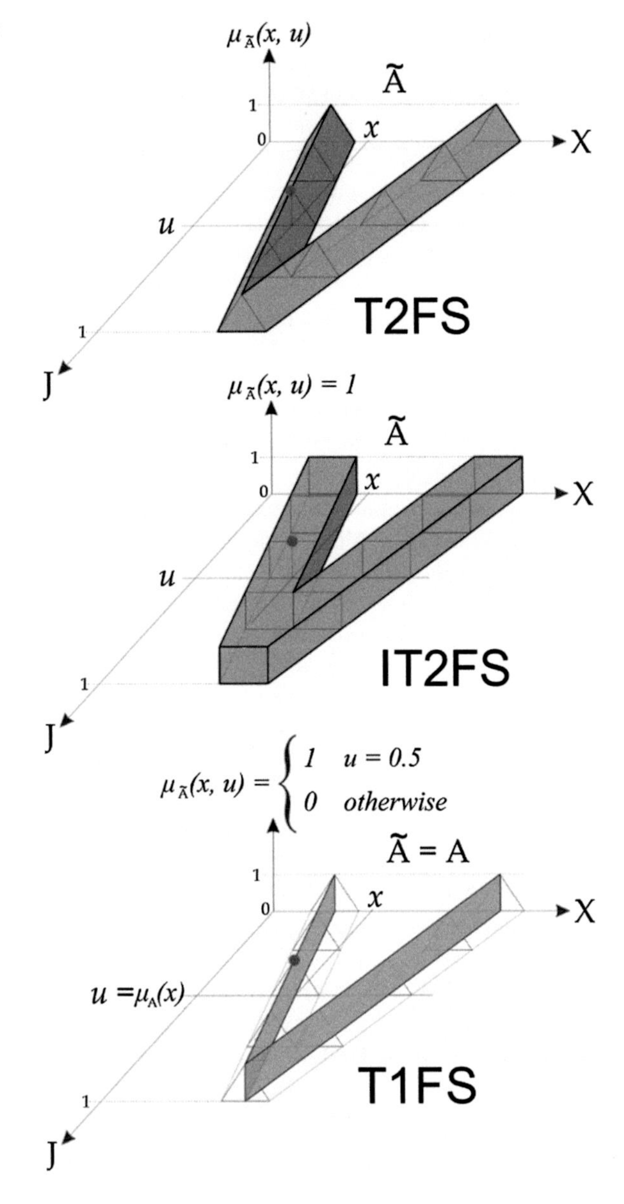

Fig. 1.23 Fuzzy type 2 a 3-D representation

$\mu_{\tilde{A}}(x, u) = 1$. In the past section, we explained the theory for T1FS which are also special cases of T2FS when the following conditions occur:

$$\mu_{\tilde{A}}(x, u) = \begin{Bmatrix} 1 & \text{if } u = 0.5 \\ 0 & \text{otherwise} \end{Bmatrix}$$

$$u = \mu_{\tilde{A}}(x)$$

1.11.3 Footprint of Uncertainty

The footprint of uncertainty (FOU) is the union of all the primary membership function and the bounded region represents the uncertainty in the primary membership function of a T2FS. There are upper and lower membership functions that are the bounds of the FOU.

The FOU can be represented by:

$$\mathrm{FOU}(\tilde{A}) = \bigcup_{x \in X} J_x \tag{1.22}$$

where the upper and lower membership functions are two type 1 fuzzy sets which bound the T2FS (see Fig. 1.24). In most of the literature are noted as: $\bar{\mu}_{\tilde{A}}(x)$ and $\underline{\mu}_{\tilde{A}}(x)$ for the upper membership function (UMF) and the lower membership function (LMF), respectively.

$$\bar{\mu}_{\tilde{A}}(x) = \max(J_x) \quad \forall x \in X \tag{1.23}$$

$$\underline{\mu}_{\tilde{A}}(x) = \min(J_x) \quad \forall x \in X \tag{1.24}$$

1.12 Fuzzy Sets Type 2 Representations

1.12.1 Digital and Continuous Representation

Similar to the type 1 fuzzy sets, the fuzzy sets type 2 can be represented in discrete $(\sum)$ or continues $(\int)$ representation.

$$\tilde{A} = \int_{x \in X} \int_{u \in J_x} \frac{\mu_{\tilde{A}}(x, u)}{(x, u)} \quad J_x \subseteq [0, 1] \tag{1.25}$$

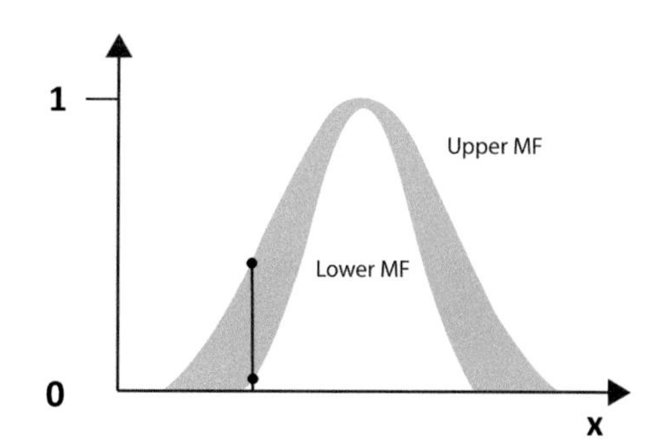

Fig. 1.24 Fuzzy type 2 membership function, the gray area is the FOU

where the $\int$ or $\sum$ operator means union of all the elements within the rank of x and u.

And Jx is the primary membership of x, the secondary membership correspond to each primary membership value and have also an interval between 0 and 1. This additional dimension gives to F2FS the possibility of having additional fuzzy sets for each value of the function and allows a great degree of uncertainty within a system. In Fig. 1.15, the membership function does not have a single value for a specific x value as T1FS; instead, each value intersects vertically many values of the degree of membership and an amplitude distribution can be assigned for every point generation a three-dimensional membership function defined for all $x \in X$. The type 2 fuzzy sets can be represented by a three-dimensional shape as can be seen in Fig. 1.14.

Vertical-slice representation

Another important representation of T2FS is the vertical-slice representation which is used for the computation of general case T2FS. In this kind of representation, the T2FS is divided in n vertical slides where each slide can be seen as a T2FS for each value of x. The membership function of each slide can be represented as follows:

$$\mu_{\tilde{A}}(x') = \int\limits_{u \in J_{x'}} \frac{f_{x'}(u)}{u} \tag{1.26}$$

where $x' \in X$ is a defined value within the rank X.

And the set $\tilde{A}$ is the union of all the vertical slices:

$$\tilde{A} = \int\limits_{x \in X} \frac{\mu_{\tilde{A}}(x)}{x} = \int\limits_{x \in X} \frac{\int\limits_{u \in J_x} \frac{f_{x'}(u)}{u}}{x} \tag{1.27}$$

Embedded representation.

Another useful representation is the embedded one, where a collection of several T1FS or T2FS within the FOU can be embedded into a system. The embedded set represents a single or a collective perception of a vague concept or idea, that is, where the opinion of many experts defer and all of them are contributing to the knowledge database.

The collective perception as an embedded system can be defined as:

$$\tilde{A}_e = \sum_{i=1}^{N} \frac{\frac{f_{x_i}(\theta_i)}{\theta_i}}{x} \quad \forall \theta \in J_x \tag{1.28}$$

where θ_i represents an embedded T1FS.

The T1FS which could conform the entire system are defined as:

$$A_e = \sum_{x \in X} \frac{\theta_i}{x} \quad \forall \theta \in J_x \tag{1.29}$$

where A_e represents only an idea of a single expert.

Operations on type 2 fuzzy sets

The theoretic operation of type 2 fuzzy sets are based on the Zadeh's extension principle [51]. Considering two T2FS $\tilde{A}$ and $\tilde{B}$ defined in a universe X with the associated membership functions $\mu_{\tilde{A}}(x)$ and $\mu_{\tilde{B}}(x)$ defined in $J_x \subseteq [0,1]$ each set is represented as $\mu_{\tilde{A}}(x) = \sum_i \frac{f_x(u_i)}{u_i}$ and $\mu_{\tilde{B}}(x) = \sum_j \frac{g_x(\omega_j)}{\omega_j}$ where $u_i, \omega_j \epsilon J$.

Using the extension principle, the membership grades for union, intersection, and negations of T2FS $\tilde{A}$ and $\tilde{B}$ can be defined as [68]:

Union:

$$\tilde{A} \cup \tilde{B} \Leftrightarrow \tilde{\mu}_{\tilde{A}\cup\tilde{B}}(x) = \tilde{\mu}_{\tilde{A}}(x) \sqcup \tilde{\mu}_{\tilde{B}}(x) = \sum_{i,j} \frac{\left(f_x(u_i) * g_x(\omega_j)\right)}{u_i \vee \omega_j} \tag{1.30}$$

Intersection:

$$\tilde{A} \cap \tilde{B} \Leftrightarrow \tilde{\mu}_{\tilde{A}\cap\tilde{B}}(x) = \tilde{\mu}_{\tilde{A}}(x) \sqcap \tilde{\mu}_{\tilde{B}}(x) = \sum_{i,j} \frac{\left(f_x(u_i) * g_x(\omega_j)\right)}{u_i * \omega_j} \tag{1.31}$$

Complement:

$$\bar{\tilde{A}} \Leftrightarrow \tilde{\mu}_{\bar{\tilde{A}}}(x) = \neg\tilde{\mu}_{\bar{\tilde{A}}}(x) = \sum_i \frac{f_x(u_i)}{i - u_i} \tag{1.32}$$

where $\vee$ represent the t-conorm (max) and $*$ represents the t-norm (min) and the summations indicate union on discrete time definition.

Join and meet under t-Norm (min)

Those operations are defined by the next theorem [68]:

Theorem 1 *Suppose that we have two convex, normal, type 1 real fuzzy sets $\tilde{F}$ and $\tilde{G}$ characterized by membership functions f and g, respectively. Let $v_0 \in \Re$ and $v_1 \in \Re$ be such that $f(v_0) = g(v_1) = 1$. Then the membership functions of the join and meet of $\tilde{F}$ and $\tilde{G}$ using max t-conorm and min t norm can be expressed as:*

$$\mu_{\tilde{F}\sqcup\tilde{G}}(\theta) = f(\theta) \wedge g(\theta); \theta \langle v_0 = g(\theta); v_0 \leq \theta \leq v_1 = f(\theta) \vee g(\theta); \theta \rangle v_1 \tag{1.33}$$

and

$$\mu_{\tilde{F}\sqcap\tilde{G}}(\theta) = f(\theta) \vee g(\theta); \theta\langle v_0 = f(\theta); v_0 \leq \theta \leq v_1 = f(\theta) \wedge g(\theta); \theta\rangle v_1 \quad (1.34)$$

Join and meet under product t-Norm

Theorem 2 Suppose that we have two convex, normal type 1 real fuzzy sets $\tilde{F}$ and $\tilde{G}$ characterized by membership functions f and g, respectively. Let $v_0 \in \Re$ and $v_1 \in \Re$ be such that $v_0 \leq v_1$ and $f(v_0) = g(v_1) = 1$. Then the membership functions of the joint $\tilde{F}$ and $\tilde{G}$, using max t-conorm and product t-norm, can be expressed as:

$$\mu_{\tilde{F}\sqcup\tilde{G}}(\theta) = f(\theta)g(\theta); \theta\langle v_0 = g(\theta); v_0 \leq \theta \leq v_1 = f(\theta) \vee g(\theta); \theta\rangle v_1 \quad (1.35)$$

The meet operation under product t-norm is not well defined but can be approximated by the following expression:

If there are n Gaussian fuzzy sets $\tilde{F}_1$, $\tilde{F}_2$, …, $\tilde{F}_n$ with means $m_1, m_2, \ldots, m_n$ and standard deviations $\sigma_1, \sigma_1, \ldots, \sigma_n$ respectively then

$$\mu_{\tilde{F}_1\sqcap\tilde{F}_2\sqcap\cdots\sqcap\tilde{F}_n}(\theta) \approx e^{-\frac{1}{2}\left(\frac{\theta - m_1 m_2 \ldots m_n}{\overline{\sigma}}\right)^2} \quad (1.36)$$

where $(i = 1, \ldots, n)$

$$\overline{\sigma} = \sqrt{\sigma_1^2 \prod_{i;i\neq 1} m_i^2 + \cdots + \sigma_j^2 \prod_{i;i\neq j} m_i^2 + \cdots + \sigma_n^2 \prod_{i;i\neq n} m_i^2} \quad (1.37)$$

This approximation is applicable for Gaussian membership functions.

Negation

Therorem 3 If a type 1 fuzzy set $\tilde{F}$ has a membership function $f(v)(v \in \Re)$, $\neg\tilde{F}$ has a membership function $f(v)(v \in \Re)$.

The joint, meet, and negation operation can be performed between membership functions of type 2 sets if unions, intersections and complements the set exists.

1.13 Interval Type 2 Fuzzy Sets (IT2FS)

In spite of the advantages of having fuzzy a third dimension which defines the degree of membership for every point within a vertical line of the FOU, the computation required for the general case makes T2FS unenforceable because for general T2FS is prohibitive to calculate meet operations for each fired rule especially if the product t-norm is used [54] due to the immense amount of computational effort required.

For solving this problem, Gorzałczany [57] and latter Mendel [61] proposed a simple method called interval type 2 fuzzy set (IT2FS) for fast computation applications. IT2FS are a particular case where only the value of 1 defines the secondary membership function.

As mentioned previously, IT2FS is a special case where $\mu_{\tilde{A}}(x, u) = 1$ and therefore the definition in the integral form is:

$$\tilde{A} = \int_{x \in X} \int_{u \in J_x} \frac{1}{(x, u)} \quad J_x \subseteq [0, 1] \tag{1.38}$$

The computation requirements for this method are much lower that for the general case. The shape of this function is also three dimensional but the secondary membership value is always 1.

As general case fuzzy sets can be represented as a vertical-slice as follows:

$$\mu_{\tilde{A}}(x') = \int_{u \in J_{x'}} \frac{1}{u} \tag{1.39}$$

$$\tilde{A} = \int_{x \in X} \frac{\mu_{\tilde{A}}(x)}{x} = \int_{x \in X} \frac{\int_{u \in J_{x'}} \frac{1}{u}}{x} \tag{1.40}$$

And for the embedded set representation as:

$$\tilde{A}_e = \sum_{i=1}^{N} \frac{\left[\frac{1}{\theta_i}\right]}{x} \quad \forall \theta \in J_x \tag{1.41}$$

where each individual T1FS can be represented as:

$$A_e = \sum_{x \in X} \frac{\theta_i}{x} \quad \forall \theta \in J_x \tag{1.42}$$

For IT2FS, the computation time gets reduced dramatically but on the other hand the potential of embedding a fuzzy system within another fuzzy system is lost; however, it has been demonstrated [54] in many applications that the IT2FS have a better performance that T1FS for noisy and not well-defined systems. That is, although all the secondary fuzzy sets have the same weight, the implementation can handle uncertainties of the system.

1.14 Type Reduction and Defuzzification

Type reduction are extended versions of type 1 defuzzification methods. In T1FS, each fired rule determines the corresponding crisp value of the output. The defuzzifier somehow combines the output sets corresponding to all the fired rules to obtain a single output set which is equivalent to have a T1FS.

As in T1FS, the defuzzification consist of obtaining a crisp value based on the position of the centroid of all the activated output sets. Using the extension principle, the centroid can be obtained by the union of all the activated sets within a defined domain. The calculation of the centroid of a type 2 set leads to a type 1 set so it is called a "type reduced" set.

For a T1FS the centroid, if the domain is discretized in N pints can be founded as:

$$C_A = \frac{\sum_{i=1}^{N} x_i \mu_A(x_i)}{\sum_{i=1}^{N} \mu_A(x_i)} \tag{1.43}$$

And for a T2FS $\tilde{A}$ the output can be obtained using the extension principle:

$$C_{\tilde{A}} = \frac{\int_{\theta_1} \cdots \int_{\theta_N} [\mu_{D1}(\theta_1) * \cdots * \mu_{DN}(\theta_N)}{\sum_{i=1}^{N} x_i\theta_i / \sum_{i=1}^{N} \theta_i} \tag{1.44}$$

where $\theta_i \in D_i$ and $*$ represents the t-norm used (normally min). Each point x_i of the $\tilde{A}$ has a type 1 membership degree $D_i = \mu_{\tilde{A}}(x_i)$, for finding the centroid all the combinations from θ_i to θ_N are evaluated and a type 1 centroid calculation is computed assigned a certain degree of membership equal to the t-norm of the membership grade.

If the domain of $\tilde{A}$ is continuous, the number of fuzzy sets embedded is infinite; therefore, it is necessary to discretize the calculation of $C_{\tilde{A}}$ using the minimum t norm for the centroid calculation. If the domain is discretized onto M points and the possible combinations of thetas is N, the number of operations required is M^N, which is a large number even for small numbers of N and M, so many optimization algorithms have been developed in order to minimize the number of operations.

Next is shown the most important defuzzification methods.

1.14.1 Karnik–Mendel Iterative Procedure (KM)

This algorithm is the most widely adopted method for type-reducing and IT2FS [52]. The type reduced new set is a special case of a T1FS called interval set. The iterative procedure is efficient to find the endpoint of the interval, as there is an element of approximation in the defuzzified value located at the midpoint. The left and right centroids are founded first and then the center point is obtained by the average of the obtained centroids.

For this algorithm, the pseudo-program for finding the left centroid c_l is the following:

1. Sort all the discourse universe values y_i in ascending order, where $i = 1, 2, \ldots, N$, such $y_1 \leq y_2 \leq \ldots \leq y_N$. Associate each y_i with its corresponding $\underline{\mu}_{\tilde{B}}(y_i)$ and $\bar{\mu}_{\tilde{B}}(y_i)$
2. Initialize

$$\theta_i = \frac{\tilde{\mu}_{\tilde{B}}(y_i) + \bar{\mu}_{\tilde{B}}(y_i)}{2}$$

3. Compute

$$y = \frac{\sum_{i=1}^{N} y_i \theta_i}{\sum_{i=1}^{N} \theta_i}$$

4. Find the switch point k, such $1 \leq k \leq N - 1$ and $y_k \leq y \leq y_{k+1}$
5. Establish

$$\theta_i = \begin{cases} \bar{\mu}_{\tilde{B}}(y_i) & i \leq k \\ \underline{\mu}_{\tilde{B}}(y_i) & \text{otherwise} \end{cases}$$

6. Compute

$$y' = \frac{\sum_{i=1}^{N} y_i \theta_i}{\sum_{i=1}^{N} \theta_i}$$

7. Check if $y' = y$. If true, stop and assign $c_l = y$, else continue.
8. Assign $y = y'$ y go to step 4.

The next step is to find the right centroid c_l and the procedure is the following:

1. Sort all the discourse universe values y_i in ascending order, where $i = 1, 2, \ldots, N$, such $y_1 \leq y_2 \leq \ldots \leq y_N$. Associate each y_i with its corresponding $\underline{\mu}_{\tilde{B}}(y_i)$ and $\bar{\mu}_{\tilde{B}}(y_i)$.
2. Initialize

$$\theta_i = \frac{\underline{\mu}_{\tilde{B}}(y_i) + \bar{\mu}_{\tilde{B}}(y_i)}{2}$$

3. Compute

$$y = \frac{\sum_{i=1}^{N} y_i \theta_i}{\sum_{i=1}^{N} \theta_i}$$

4. Find the switch point k, such $1 \leq k \leq N - 1$ and $y_k \leq y \leq y_{k+1}$

5. Establish

$$\theta_i = \begin{cases} \underline{\mu}_{\tilde{B}}(y_i) & i \leq k \\ \bar{\mu}_{\tilde{B}}(y_i) & \text{otherwise} \end{cases}$$

6. Compute

$$y' = \frac{\sum_{i=1}^{N} y_i \theta_i}{\sum_{i=1}^{N} \theta_i}$$

7. Check if $y' = y$. If true, stop and assign $c_r = y$, else continue.
8. Assign $y = y'$ y go to step 4.

Once we obtain both centroids the general centroid can be obtained averaging the results:

$$y = \frac{c_l + c_r}{2} \tag{1.45}$$

1.14.2 *Wu-Mendel Uncertain Bounds*

In [62] is presented a better proposal that the original Karnik–Mendel first type reducer algorithm which replace type reducer with lower and upper bounds (uncertainty bounds) for the end points of the type reducer set, and those bounds, which are optimal in a minimum–maximum sense, can be computed without performing type reduction [47].

To start, four centroids or boundary T1FS are defined. All of them can be computed once the left and right points of the activated interval have been computed. yl^i and yr^i are defined as the left and right end points of the centroid of the sth consequence of the IT2FS, previously reordered in ascending order. The centroids of the boundary T1FS are:

$$\{\text{LMFs, Left}\} : y_l^{(0)}(x) = \frac{\sum_{i=1}^{M} \underline{f}^i y_l^i}{\sum_{i=1}^{M} \underline{f}^i} \tag{1.46}$$

$$\{\text{LMFs, Right}\} : y_r^{(M)}(x) = \frac{\sum_{i=1}^{M} \underline{f}^i y_r^i}{\sum_{i=1}^{M} \underline{f}^i} \tag{1.47}$$

$$\{\text{UMFs, Left}\} : y_l^{(M)}(x) \frac{\sum_{i=1}^{M} \bar{f}^i y_l^i}{\sum_{i=1}^{M} \bar{f}^i} \tag{1.48}$$

$$\{\text{LMFs, Right}\} : y_r^{(0)}(x) \frac{\sum_{i=1}^{M} \bar{f}^i y_r^i}{\sum_{i=1}^{M} \bar{f}^i} \tag{1.49}$$

According to the theorem 3.4 of [47], the endpoints $y_l(x)$ and $y_r(x)$ of the type reduced set of an interval type 2 fuzzy logic controller for the input x, are bounded from below and above as: $\underline{y_l}(x) \leq y_{l(x)} \leq \bar{y}_{l(x)}$ and $\underline{y_r}(x) \leq y_{r(x)} \leq \bar{y}_{r(x)}$where:

$$\overline{y_l} = \min \text{Error! Bookmark not defined.}$$

And

$$\underline{y_r} = \max \text{Error! Bookmark not defined.}$$

Then:

$$\underline{y_l} = \overline{y_l} - \left[\frac{\sum_{i=1}^{M} \left(\bar{f}^i - \underline{f}^i \right)}{\sum_{i=1}^{M} \bar{f}^i \sum_{i=1}^{M} \underline{f}^i} \times \frac{\sum_{i=1}^{M} \underline{f}^i \left(y_l^i - y_l^1 \right) \sum_{i=1}^{M} \bar{f}^i \left(y_l^M - y_l^i \right)}{\sum_{i=1}^{M} \underline{f}^i \left(y_l^i - y_l^1 \right) + \sum_{i=1}^{M} \bar{f}^i \left(y_l^M - y_l^i \right)} \right] \tag{1.50}$$

$$\overline{y_r} = \underline{y_r} + \left[\frac{\sum_{i=1}^{M} \left(\bar{f}^i - \underline{f}^i \right)}{\sum_{i=1}^{M} \bar{f}^i \sum_{i=1}^{M} \underline{f}^i} \times \frac{\sum_{i=1}^{M} \bar{f}^i \left(y_r^i - y_r^1 \right) \sum_{i=1}^{M} \underline{f}^i \left(y_r^M - y_r^i \right)}{\sum_{i=1}^{M} \bar{f}^i \left(y_r^i - y_r^1 \right) + \sum_{i=1}^{M} \underline{f}^i \left(y_r^M - y_r^i \right)} \right] \tag{1.51}$$

And the defuzzification is obtained averaging the four centroids obtained:

$$y = \frac{1}{4} \left(\underline{y_l} + \overline{y_l} + \underline{y_r} + \overline{y_r} \right) \tag{1.52}$$

1.14.3 Enhanced Karnik–Mendel Algorithm

This algorithms was developed and published by Wu and Mendel in [65]. In order to reduce computational cost of the original algorithm a better initializations is proposed and the end conditions of the iterations is modified in order to avoid unnecessary iterations. As the original algorithm, the left and right parts y_l and y_r needs to be calculated and then the average of both is the defuzzified crisp value.

The optimal initial switch point for the first iteration can be expressed as:

$$y_l = \frac{\sum_{i=1}^{N} \underline{x_i} \left[\frac{\overline{w_i} + \underline{w_i}}{2} \right]}{\sum_{i=1}^{N} \left[\frac{\overline{w_i} + \underline{w_i}}{2} \right]} \tag{1.53}$$

where $\overline{w_i}$ the match is weight of the respective x_i (the upper membership function value of the FOU) and $\underline{w_i}$ is the match weigh of the lower membership function according to Fig. 1.25.

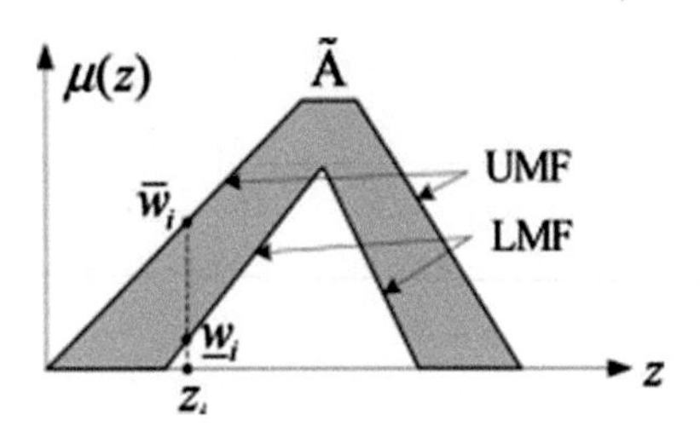

Fig. 1.25 Matched weigh for the upper and lower membership functions

When $y' = y$, it means that the following iterations is not providing any improvement to the calculated vale of y_l. If $y' = y \Rightarrow k' = k$, but the last comparison saves one iteration, so the criteria for ending the program is $k' = k$.

$$y' = \frac{\sum_{i=1}^{k'} x_i \overline{w_i} + \sum_{i=k'+1}^{N} x_i \underline{w_i}}{\sum_{i=1}^{k'} \overline{w_i} + \sum_{i=k'+1}^{N} \underline{w_i}} \tag{1.54}$$

$$y = \frac{\sum_{i=1}^{k} x_i \overline{w_i} + \sum_{i=k+1}^{N} x_i \underline{w_i}}{\sum_{i=1}^{k} \overline{w_i} + \sum_{i=k+1}^{N} \underline{w_i}} \tag{1.55}$$

Once the initial value of y_l has been calculated the algorithm of the left centroid can be computed as follow:

1. Sort all the discourse universe values $\overline{x_i}$ in ascending order, where $i = 1, 2, \ldots, N$, such that $\overline{x_1} \leq \overline{x_2} \leq \ldots \leq \overline{x_N}$. And then match the weights w_i associated with their respective x_i.
2. Establish $k = \text{round}(N/2.4)$ and compute

$$a = \sum_{i=1}^{k} y_i \bar{\mu}_{\tilde{B}}(y_i) + \sum_{i=k+1}^{N} y_i \underline{\mu}_{\tilde{B}}(y_i)$$

$$b = \sum_{i=1}^{k} \underline{\mu}_{\tilde{B}}(y_i) + \sum_{i=k+1}^{N} \underline{\mu}_{\tilde{B}}(y_i)$$

$$y = \frac{a}{b}$$

3. Find the switch point $k' \in [1, N-1]$, such that

$$y_{k\prime} \leq y \leq y_{k\prime+1}$$

4. Check if $k' = k$. If true, stop and assign $c_l = y$; else continue.
5. Compute

$$s = \text{sign}(k' - k)$$

$$a' = a + s \sum_{i=\min(\mathrm{k,k'})+1}^{\max(k,k')} y\left[\bar{\mu}_{\tilde{B}}(y_i) - \underline{\mu}_{\tilde{B}}(y_i)\right]$$

$$b' = b + s \sum_{i=\min(\mathrm{k,k'})+1}^{\max(k,k')} \left[\bar{\mu}_{\tilde{B}}(y_i) - \underline{\mu}_{\tilde{B}}(y_i)\right]$$

and compute again

$$y' = \frac{a'}{b'}$$

6. Assign $y = y', a = a', b = b', k = k\prime$ and go to step 3.

1.14.4 Type 2 Fuzzy Logic Systems Block Diagram

Figure 1.26 presents the block diagram for a T2FS which is similar to a traditional T1FS and is composed with those steps:

Fuzzifier: The fuzzifier maps the read value into the T2DS which activates the inference system.

Rule base: The antecedents and consequents are related each other by IF-ELSE language functions.

Inference: This block assign to the fuzzy input a fuzzy output according to the rules established and operators such as joint (⊔) and meet operators (⊓), those operators are equivalent to the union and intersection operations but are used in the secondary membership functions. The definition and explanation can be found in [69].

Type reduction: For some systems it is required to transform the type 2 fuzzy outputs from the inference engine into T1FS and the result is called a type reduced set. There most common method for doing this are the Karnik–Mendel iteration algorithm and the Wu-Mendel uncertainty bounds method. Both are based on the calculation of centroid of mass.

Defuzzification: Once the outputs have been reduced the defuzzification block determines the crisp value that will be introduced to the actuator. In some

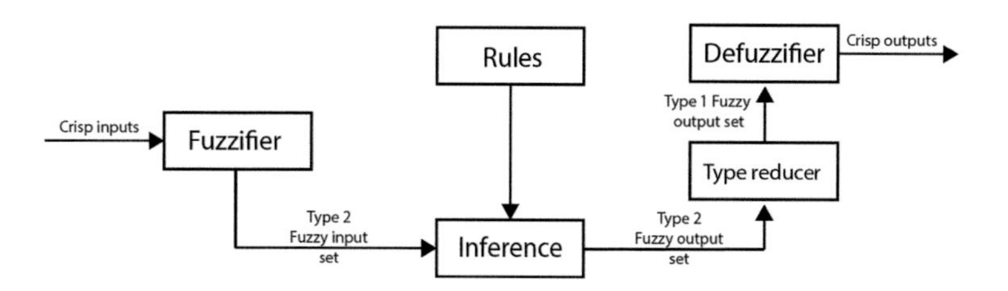

Fig. 1.26 Block diagram of a type 2 fuzzy logic system

applications, where there is not a type reducer block the defuzzification is made directly from the output of the inference engine.

1.14.5 Interval Type 2 Fuzzy Logic Numeric Example

A micro-manufacturing process is controlled by an IT2FS controller. The position error and position changes are used for calculate the voltage of a servomotor in order to modify its cutting speed. The controller algorithm structure is presented in Fig. 1.27.

INPUT 1: Where the **Position Error** has the next representation (see Fig. 1.28).

$$\overline{\mu_N} = \left\{\frac{1}{-0.025} + \frac{1}{-0.02} + \frac{1}{-0.015} + \frac{0.92}{-0.01} + \frac{0.78}{-0.005} + \frac{0.61}{0} + \frac{0.46}{0.005} + \frac{0.3}{0.01} + \frac{0.15}{0.015} + \frac{0}{0.02} + \frac{0}{0.025}\right\}$$

$$\underline{\mu_N} = \left\{\frac{1}{-0.025} + \frac{1}{-0.02} + \frac{1}{-0.015} + \frac{0.84}{-0.01} + \frac{0.57}{-0.005} + \frac{0.29}{0} + \frac{0}{0.005} + \frac{0}{0.01} + \frac{0}{0.015} + \frac{0}{0.02} + \frac{0}{0.025}\right\}$$

$$\overline{\mu_Z} = \left\{\frac{0}{-0.025} + \frac{0}{-0.02} + \frac{0.25}{-0.015} + \frac{0.5}{-0.01} + \frac{0.75}{-0.005} + \frac{1}{0} + \frac{0.75}{0.005} + \frac{0.5}{0.01} + \frac{0.25}{0.015} + \frac{0}{0.02} + \frac{0}{0.025}\right\}$$

$$\underline{\mu_Z} = \left\{\frac{0}{-0.025} + \frac{0}{-0.02} + \frac{0}{-0.015} + \frac{0}{-0.01} + \frac{0}{-0.005} + \frac{1}{0} + \frac{0}{0.005} + \frac{0}{0.01} + \frac{0}{0.015} + \frac{0}{0.02} + \frac{0}{0.025}\right\}$$

$$\overline{\mu_P} = \left\{\frac{0}{-0.025} + \frac{0}{-0.02} + \frac{0.15}{-0.015} + \frac{0.3}{-0.01} + \frac{0.46}{-0.005} + \frac{0.61}{0} + \frac{0.78}{0.005} + \frac{0.92}{0.01} + \frac{1}{0.015} + \frac{1}{0.02} + \frac{1}{0.025}\right\}$$

$$\underline{\mu_P} = \left\{\frac{0}{-0.025} + \frac{0}{-0.02} + \frac{0}{-0.015} + \frac{0}{-0.01} + \frac{0}{-0.005} + \frac{0.29}{0} + \frac{0.57}{0.005} + \frac{0.84}{0.01} + \frac{1}{0.015} + \frac{1}{0.02} + \frac{1}{0.025}\right\}$$

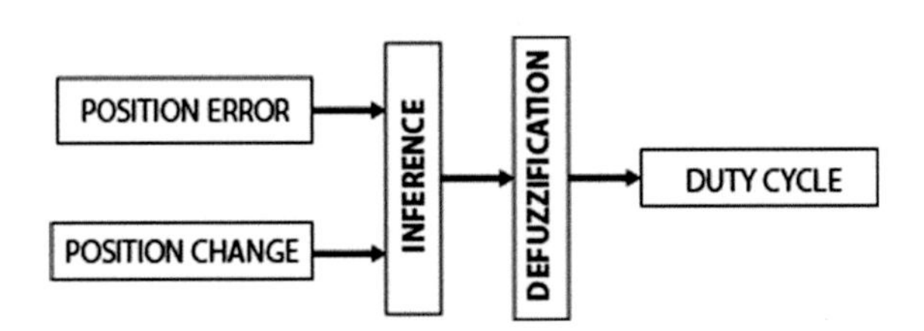

Fig. 1.27 Fuzzy logic control type 2 representation

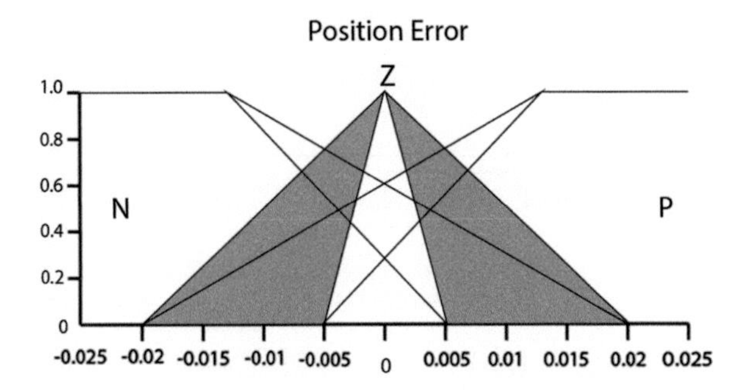

Fig. 1.28 Input variable ERROR

INPUT 2: Where the **Position Change** has the next representation (see Fig. 1.29).

$$\overline{\mu_N} = \left\{ \frac{1}{-0.05} + \frac{1}{-0.04} + \frac{1}{-0.03} + \frac{0.91}{-0.02} + \frac{0.75}{-0.01} + \frac{0.59}{0} + \frac{0.41}{0.01} + \frac{0.25}{0.02} + \frac{0.08}{0.03} + \frac{0}{0.04} + \frac{0}{0.05} \right\}$$

$$\underline{\mu_N} = \left\{ \frac{1}{-0.05} + \frac{1}{-0.04} + \frac{1}{-0.03} + \frac{0.87}{-0.02} + \frac{0.61}{-0.01} + \frac{0.38}{0} + \frac{0.12}{0.01} + \frac{0}{0.02} + \frac{0}{0.03} + \frac{0}{0.04} + \frac{0}{0.05} \right\}$$

$$\overline{\mu_Z} = \left\{ \frac{0}{-0.05} + \frac{0}{-0.04} + \frac{0.15}{-0.03} + \frac{0.44}{-0.02} + \frac{0.72}{-0.01} + \frac{1}{0} + \frac{0.72}{0.01} + \frac{0.44}{0.02} + \frac{0.15}{0.03} + \frac{0}{0.04} + \frac{0}{0.05} \right\}$$

$$\underline{\mu_Z} = \left\{ \frac{0}{-0.05} + \frac{0}{-0.04} + \frac{0}{-0.03} + \frac{0}{-0.02} + \frac{0.33}{-0.01} + \frac{1}{0} + \frac{0.33}{0.01} + \frac{0}{0.02} + \frac{0}{0.03} + \frac{0}{0.04} + \frac{0}{0.05} \right\}$$

$$\overline{\mu_P} = \left\{ \frac{0}{-0.05} + \frac{0}{-0.04} + \frac{0.08}{-0.03} + \frac{0.25}{-0.02} + \frac{0.41}{-0.01} + \frac{0.59}{0} + \frac{0.75}{0.01} + \frac{0.91}{0.02} + \frac{1}{0.03} + \frac{1}{0.04} + \frac{1}{0.05} \right\}$$

$$\underline{\mu_P} = \left\{ \frac{0}{-0.05} + \frac{0}{-0.04} + \frac{0}{-0.03} + \frac{0}{-0.02} + \frac{0.12}{-0.01} + \frac{0.38}{0} + \frac{0.61}{0.01} + \frac{0.87}{0.02} + \frac{1}{0.03} + \frac{1}{0.04} + \frac{1}{0.05} \right\}$$

Using the two inputs described, it is possible to define the following rules.

IF Position Error is Negative AND Position Change is Positive THEN the output is Negative

Rule	P. Error	AND	P. Change	Output
1	*N*	∩	*N*	*N*
2	*N*	∩	*Z*	*N*
3	*N*	∩	*P*	*N*
4	*Z*	∩	*N*	*N*
5	*Z*	∩	*Z*	*Z*
6	*Z*	∩	*P*	*P*
7	*P*	∩	*N*	*P*
8	*P*	∩	*Z*	*P*
9	*P*	∩	*P*	*P*

The parameters of the output (Voltage) are (see Fig. 1.30).

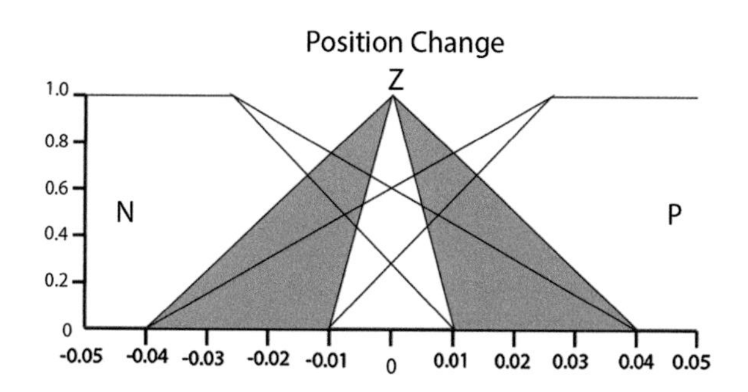

Fig. 1.29 Input variable position change

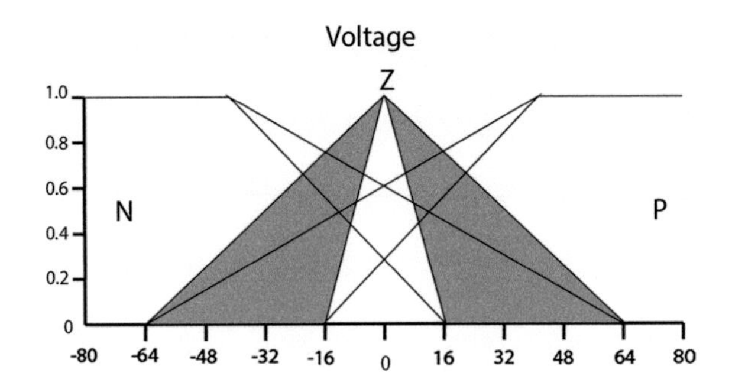

Fig. 1.30 Input variable position change

$$\overline{\mu_N} = \left\{\frac{1}{-80}+\frac{1}{-64}+\frac{1}{-48}+\frac{0.86}{-32}+\frac{0.58}{-16}+\frac{0.29}{0}+\frac{0}{16}+\frac{0}{32}+\frac{0}{48}+\frac{0}{64}+\frac{0}{80}\right\}$$

$$\underline{\mu_N} = \left\{\frac{1}{-80}+\frac{1}{-64}+\frac{1}{-48}+\frac{0.66}{-32}+\frac{0}{-16}+\frac{0}{0}+\frac{0}{16}+\frac{0}{32}+\frac{0}{48}+\frac{0}{64}+\frac{0}{80}\right\}$$

$$\overline{\mu_Z} = \left\{\frac{0}{-80}+\frac{0}{-64}+\frac{0.15}{-48}+\frac{0.43}{-32}+\frac{0.71}{-16}+\frac{1}{0}+\frac{0.71}{16}+\frac{0.43}{32}+\frac{0.15}{48}+\frac{0}{64}+\frac{0}{80}\right\}$$

$$\underline{\mu_Z} = \left\{\frac{0}{-80}+\frac{0}{-64}+\frac{0}{-48}+\frac{0}{-32}+\frac{0.33}{-16}+\frac{1}{0}+\frac{0.33}{16}+\frac{0}{32}+\frac{0}{48}+\frac{0}{64}+\frac{0}{80}\right\}$$

$$\overline{\mu_P} = \left\{\frac{0}{-80}+\frac{0}{-64}+\frac{0}{-48}+\frac{0}{-32}+\frac{0}{-16}+\frac{0.29}{0}+\frac{0.58}{16}+\frac{0.86}{32}+\frac{1}{48}+\frac{1}{64}+\frac{1}{80}\right\}$$

$$\underline{\mu_P} = \left\{\frac{0}{-80}+\frac{0}{-64}+\frac{0}{-48}+\frac{0}{-32}+\frac{0}{-16}+\frac{0}{0}+\frac{0}{16}+\frac{0.66}{32}+\frac{1}{48}+\frac{1}{64}+\frac{1}{80}\right\}$$

As a numerical example, it is evaluated the Position error in −0.001 and Position change in 0.001 obtaining with the rules, and applying the minimum T-Norm the next results are calculated:

Rule	P. Error	AND	P. Change	Minimum	Output
1	*N* (0.92/0.84)	∩	*N* (0.41/0.12)	Min((0.92/0.84), (0.41/0.12))	(0.41/0.12)
2	*N* (0.92/0.84)	∩	*Z* (0.72/0.33)	Min((0.92/0.84), (0.72/0.33))	(0.72/0.33)
3	*N* (0.92/0.84)	∩	*P* (0.75/0.61)	Min((0.92/0.84), (0.75/0.61))	(0.75/0.61)
4	*Z*(0.5/0)	∩	*N* (0.41/0.12)	Min((0.5/0), (0.41/0.12))	(0.41/0.12)
5	*Z*(0.5/0)	∩	*Z* (0.72/0.33)	Min((0.5/0), (0.72/0.33))	(0.5/0)
6	*Z*(0.5/0)	∩	*P* (0.75/0.61)	Min((0.5/0), (0.75/0.61))	(0.5/0)
7	*P*(0.3/0)	∩	*N* (0.41/0.12)	Min((0.3/0), (0.41/0.12))	(0.3/0)

(continued)

(continued)

Rule	P. Error	AND	P. Change	Minimum	Output
8	*P*(0.3/0)	∩	*Z* (0.72/0.33)	Min((0.3/0), (0.72/0.33))	(0.3/0)
9	*P*(0.3/0)	∩	*P* (0.75/0.61)	Min((0.3/0), (0.75/0.61))	(0.3/0)

Then, the aggregation can be performed with the Maximum T-Conorm, according to the rule set, such as:

Rule	P. Error	AND	P. Change	Result	Maximum	Output
1	*N*	∩	*N*	(0.41/0.12)	Max((0.41/0.12), (0.72/0.33), (0.75/0.61), (0.41/0.12))	(0.75/0.61)
2	*N*	∩	*Z*	(0.72/0.33)		
3	*N*	∩	*P*	(0.75/0.61)		
4	*Z*	∩	*N*	(0.41/0.12)		
5	*Z*	∩	*Z*	(0.5/0)	–	(0.5/0)
6	*Z*	∩	*P*	(0.5/0)	Max((0.5/0), (0.3/0), (0.3/0), (0.3/0))	(0.5/0)
7	*P*	∩	*N*	(0.3/0)		
8	*P*	∩	*Z*	(0.3/0)		
9	*P*	∩	*P*	(0.3/0)		

With these three values, we can infer the final IT2FS from the Voltage sets (see Fig. 1.31).

$$\overline{\mu_{\text{inferred}}} = \left\{\frac{0.75}{-80} + \frac{0.75}{-64} + \frac{0.75}{-48} + \frac{0.75}{-32} + \frac{0.58}{-16} + \frac{0.5}{0} + \frac{0.5}{16} + \frac{0.5}{32} + \frac{0.5}{48} + \frac{0.5}{64} + \frac{0.5}{80}\right\}$$

$$\underline{\mu_{\text{inferred}}} = \left\{\frac{0.61}{-80} + \frac{0.61}{-64} + \frac{0.61}{-48} + \frac{0.61}{-32} + \frac{0}{-16} + \frac{0}{0} + \frac{0}{16} + \frac{0}{32} + \frac{0}{48} + \frac{0}{64} + \frac{0}{80}\right\}$$

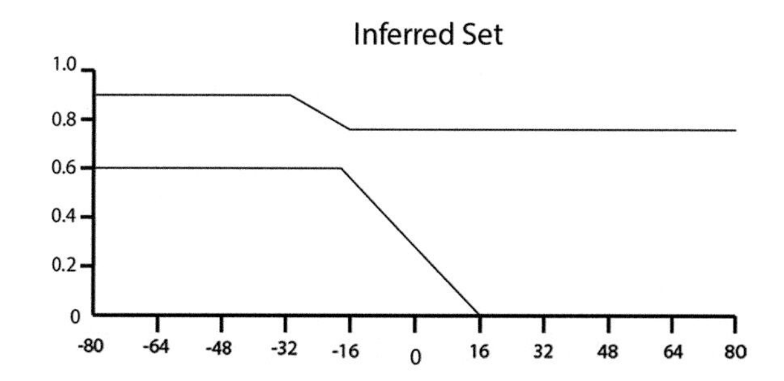

Fig. 1.31 Inferred set

Then, the output centroid is calculated, which means the voltage applied with the Nie-Tan Method:

$$Y_{Voltage} = \frac{\sum_{i=1}^{N}\left(\bar{\mu}_{Inferred}(y_i) + \underline{\mu}_{Inferred}(y_i)\right)y_i}{\sum_{i=1}^{N}\bar{\mu}_{Inferred}(y_i) + \sum_{i=1}^{N}\underline{\mu}_{Inferred}(y_i)}$$

$$Y_{DutyCycle} = \frac{(0.75+0.61)(-80)+(0.75+0.61)(-64)+(0.75+0.61)(-48)(0.75+0.61)(-32)+(0.58+0)(-16)}{(0.75+0.61)+(0.75+0.61)+(0.75+0.61)+(0.75+0.61)+(0.58+0)+(0.5+0)}$$
$$+\frac{(0.5+0)(0)+(0.5+0)(16)+(0.5+0)(32)+(0.5+0)(48)+(0.5+0)(64)+(0.5+0)(80)}{(0.5+0)+(0.5+0)+(0.5+0)+(0.5+0)+(0.5+0)}$$

$$Y_{Voltage} = -21.5\,\%$$

1.15 Experimental Implementation of a Fuzzy Logic Controller Type-2 in Quadrotors

1.15.1 Introduction

The study of fuzzy logic controllers for quadrotors have been developed highly in the last decade. Coza and Macnab presented in 2006 a new robust adaptive fuzzy control method for quadrotor stabilization. They propose a method using control and center updates for each axis rotation to approximate the same nonlinear function as the e-modification method. Results on Simulink showed that this new method offered a better center tuning than the e-method, obtaining less error in the steady state although oscillations were present in a range of ±0.2 radians [70].

In 2010 Santos et al. [71] developed a PID-like type-1 fuzzy controller for the three axes of a Quadrotor and the altitude using trapezoid membership functions. The controller was tested in Simulink showing a smooth, fast, stable response Santos [71]. The same year, Kirly et. al. presented their work of the design of a Fuzzy controller embedded into a TMS320F28335 micro processing unit, testing the axes separately. They obtained that starting from 3° and 12°, the controller was able to reach its steady state of ±2° near-horizontal and good response to perturbations. They note also that given the sensitiveness of their inclination sensor, vibrations given by environmental noises are to be studied extensively [72].

In 2013, Sheikpour and Shouraki published their results of the design of a Fuzzy controller using the Parallel Distributed Compensation method obtained from a Takagi-Sugeno fuzzy model of a Quadrotor. They show the viability of this method obtaining a ~1 s response with a ~0.1 radians of overshoot [73]. The same year, Ilhan and Karakose [74] presented their work in Type-2 Fuzzy Logic controller for a Quadrotor for position and altitude using triangular membership functions. It showed a very slightly better response than type-1 Fuzzy approach and significantly better than PID. However, it was not delved in the tuning and selection of different Footprints of Uncertainty [74].

In 2014, was presented a real-time fuzzy controller embedded in a GUMSTIX Overo FIRESTORM COM microcontroller board. The controller was obtained using the ANFIS system from test data obtained from a first experiment. The results showed that fuzzy controller is easily capable of controlling the Quadrotor, with the advantage that it was self-tuned as opposed to the PD controller. Besides, fuzzy outperformed PD in certain conditions [75]. Finally, in 2014, a Hybrid method of backstepping and fuzzy adaptive PID is proposed by Qingji et al. [76]. In it, a fuzzy inference system is used to tune the parameters Kp, Ki, and Kd of a PID controller. Simulation and practical results showed that this hybrid controller performed achieved the stabilization, effectiveness, and robustness desired, with variations of ± 1 degree in the steady state and rejection of disturbances of $\sim 5°$ [76].

1.15.2 Quadrotor Basic Principles

It is important to describe how the Quadrotor works in order to design its controller. The arrangement of motors in the Quadrotor is shown in Fig. 1.32. Each rotor has a thrust and an angular momentum about its center of rotation, as well as a drag force opposite to the rotorcraft's direction of flight [77]. To produce lift, the rotors have to spin at a certain speed to produce enough thrust. The quantity of thrust will determine the altitude and speed at which the Quadrotor rises [78]. On the other hand, the spinning of each motor generates an angular momentum that will try to rotate the Quadrotor in its yaw angle such as it would happen in a helicopter without tail propeller. To avoid this effect, two intercalated rotors will spin clockwise and the other two counterclockwise so that the individual angular

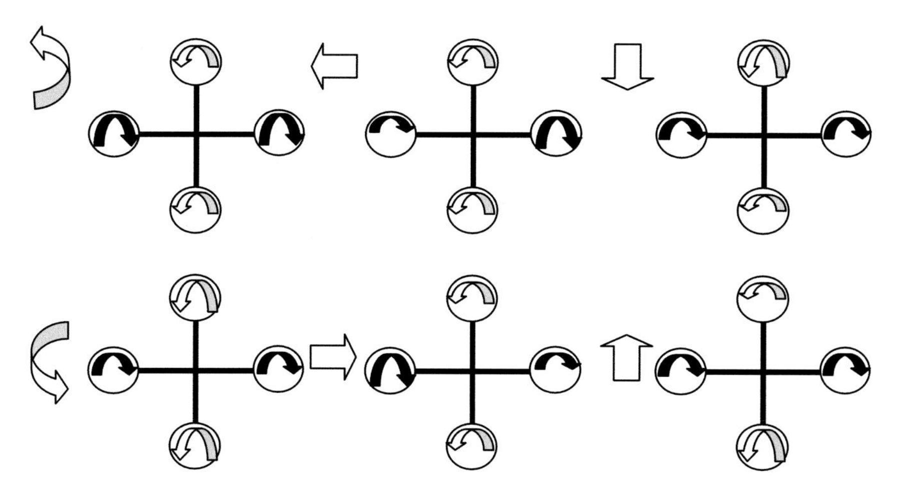

Fig. 1.32 Quadrotor frame and functioning (**a** clockwise yaw, **b** anticlockwise yaw, **c** anticlockwise roll, **d** clockwise roll, **e** anticlockwise pitch, **f** clockwise pitch)

momentum cancel each other. Therefore, the propellers attached to each motor are different: two of them are pusher and two are puller, working in contrarotation.

In order to change the angle along pitch and roll axes, and therefore move it in a certain direction, depending on the desired speed of the displacement, the rotors orientated toward that direction must change their thrust as seen in the Fig. 1.32, but always taking care that the sum of angular momentum remains zero so that the Quadrotor remains in the same altitude and yaw angle.

Basic Diagram for Fuzzy Logic Type 2 using in the Quadrotor
Figure 1.33 shows the main block diagram for a Fuzzy Logic Controller Type- 2 (T2FS), which is similar to a traditional Fuzzy Logic Type 1, presented in this chapter.

Fuzzifier: The fuzzifier maps the read value into the T2DS
which activates the inference system.

Rule base: The antecedents and consequents are related each other by IF–ELSE language functions.

Inference: This block assign to the fuzzy input a fuzzy output according to the rules established and operators such as union ($\sqcup$) and intersection operators ($\sqcap$); those operators are equivalent to the union and intersection operations but are used in the secondary membership functions. The definition and explanation can be found in the paper presented by Mendel [79].

Type reduction: For some systems, it is required to transform the type 2 fuzzy outputs from the inference engine into T1FS and the result is called a type reduced set. The most common methods for doing this are the Mendel 2007 iteration algorithm and the [79] uncertainty bounds method. Both are based on the calculation of centroid of mass. Defuzzification: once the outputs have been reduced, the defuzzification block determines the crisp value that will be send to the actuator 80.

1.15.3 ANFIS

Fuzzy logic and neural networks are complementary tools for building intelligent systems. While neural networks are low-level computational structures that perform well when dealing with raw data, fuzzy logic deals with reasoning on a higher level,

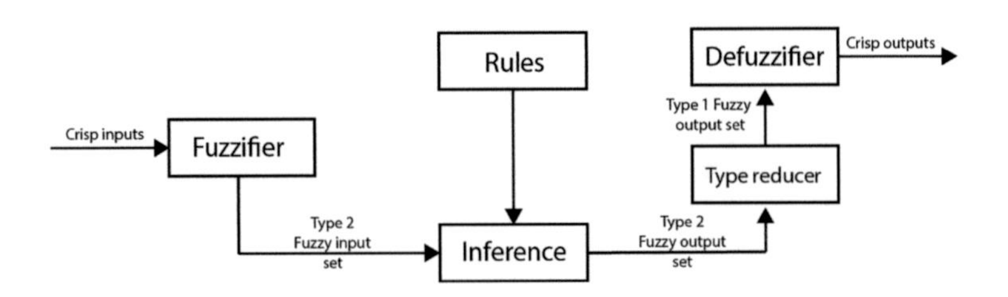

Fig. 1.33 Block diagram of a type 2 fuzzy logic system

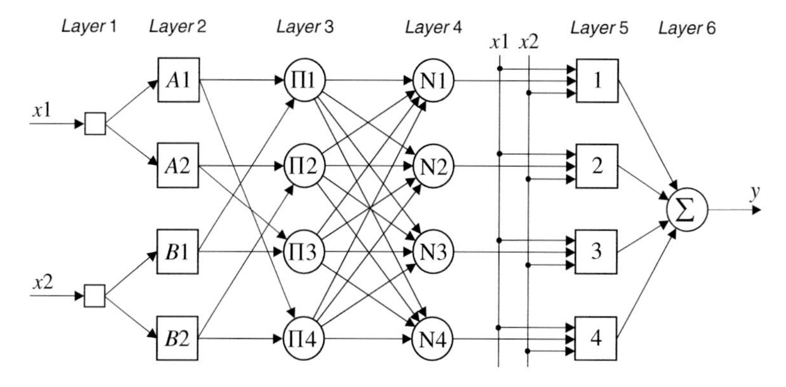

Fig. 1.34 Depicts an ANFIS topology

using linguistic information acquired from domain experts. However, fuzzy systems lack the ability to learn and cannot adjust themselves to a new environment. On the other hand, although neural networks can learn, they are black boxes to the user. When a training input–output example is presented to the system, the backpropagation algorithm can compute the system output and compares it with the desired output of the training example. The error is propagated backward through the network from the output layer to the input layer. The neuron activation functions are modified as the error is propagated. To decide the necessary modifications, the backpropagation algorithm differentiates the activation functions of the neurons. Figure 1.34 depicts an ANFIS topology [81].

1.16 Design of Fuzzy Logic Controller Tuned by an Expert

The Fuzzy Logic controller proposed in this paper is a Type-2. To produce the membership functions and inference rules for the controller; it is also fundamental to set up the control target: the Quadrotor must be able to keep the requested orientation of its three rotation angles, so that it follows an established reference threshold (±10° around the horizontal for the case of x and y axes and the initial value for the case of z), while it is capable of rejecting external physical disturbances of restrained amplitude applied over its axes. Given that the motors are 90° from each other, it can be assumed that there are two axes with one motor in each side and that the motors that belong to the same axis turn in the same direction. The sensor used is aligned in such a way that a rotation in one physical axis of the Quadrotor provides primarily changes in just one of the measured axes. With this, so as to vary the rotation in one angle (either x or y from Fig. 1.35), it is just necessary to adjust the output of the motors attached to the perpendicular axis. In the case of adjusting the rotation about z, it is necessary to increase equally the torque of the motors of one axis while reducing in the same value the ones from the other axis. This variation generates the correction in the z axis and avoids variations in x and y or the altitude.

From the gyroscope and accelerometer, it is possible to acquire the measurements of rotation angle and differentials of this value. The parameters to be considered as inputs and outputs are shown in Table 1.3.

According with all the experimentation with fuzzy logic controllers, it is determined that 7 membership functions for the rotation of each angle, and 5 for the derivatives will create a sufficient description of the system. This creates the possibility to use a larger number of functions close to the reference value, since a significant part of the control takes part over that interval. The latter implies that while there is a membership function center in zero, the next ones will be close to this value and the extreme functions will consider an angle value which the Quadrotor and the controller could not be capable to handle properly anymore generating an output that will try to correct it. The shape selected for the MFs is the Gaussian. The MFs are the symmetrical for all the axes because it is expected that they react in the same manner. The final MFs adjusted are illustrated in Figs. 1.36 and 1.37.

The Linguistic variables that describe each membership function (inputs and outputs) are illustrated in Table 1.4.

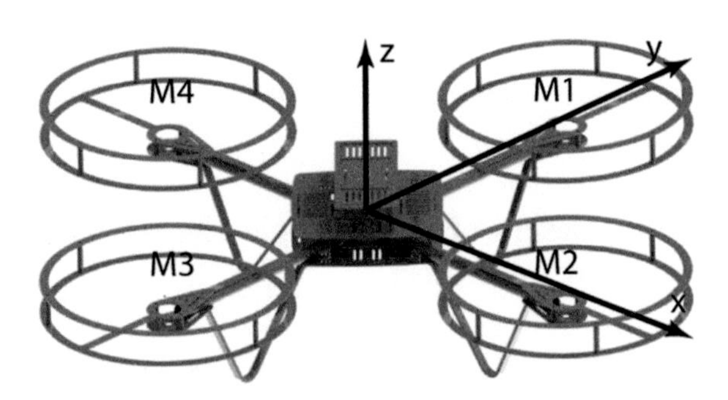

Fig. 1.35 Axes and motors in the Quadrotor frame

Table 1.3 Inputs and outputs of the system

x	Angle in the roll axis
dx	Angular velocity in the roll axis
y	Angle in the pitch axis
dy	Angular velocity in the pitch axis
z	Angle in the yaw axis
dz	Angle in the yaw axis
M1	% of PWM for Motor 1
M2	% of PWM for Motor 2
M3	% of PWM for Motor 3
M4	% of PWM for Motor 4

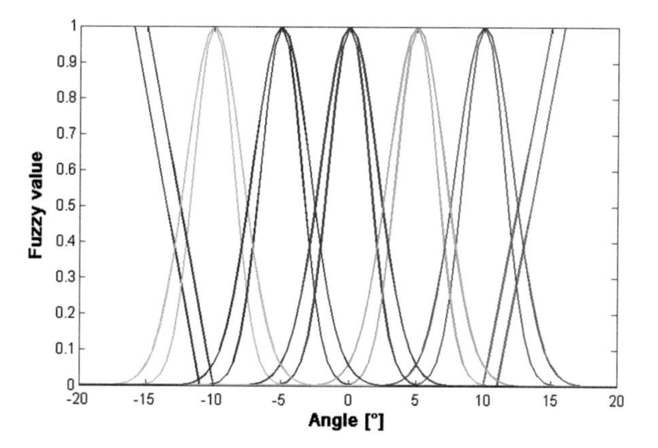

Fig. 1.36 Membership Functions for x, y and z

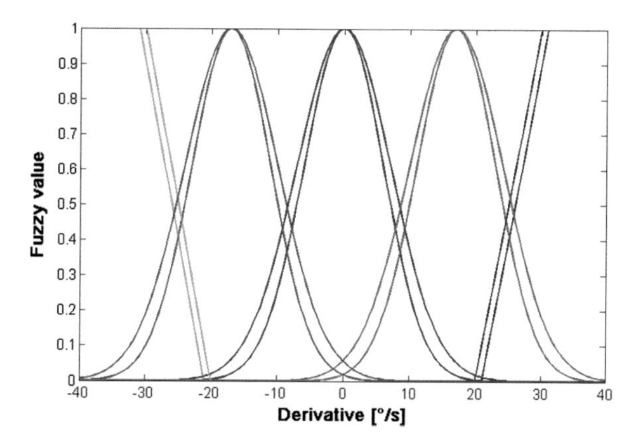

Fig. 1.37 Membership Functions for dx, dy and dz

For the output, nine singleton MFs were defined. These MFs represent the PWM percentage value for the motors. The MFs are mounted over a base value of PWM percentage, which is established to be 60. Since the base value was calibrated for each motor, the MFs are the same for the four motors. Figure 1.38 shows the final MFs of the outputs:

As a result, the physical layout of the Quadrotor explained before (See Fig. 1.35). Table 1.5 shows the mapping for the inference rules that relates the input MFs with the output MFs. All the rules addressed the general expression (if-then):

If X is A and Y is B then Z is C.

The expert can generate the linguistic rules from the effects generated in the Quadrotor by each motor; two examples are provided for the case of Motor 2 (see Fig. 1.39).

If the Quadrotor has a negative large angle (Motor 2 is below the horizontal line, Fig. 1.39) and the derivative is negative medium (it is rotating in the same direction of the negative angle), then Motor 2 needs a Positive Extra-large PWM value in

Table 1.4 Linguistic values of the MFs

NXL	Negative Extra-large
NL	Negative large
NM	Negative medium
NS	Negative small
Z	Zero
PS	Positive small
PM	Positive medium
PL	Positive large
PXL	Positive extra-large

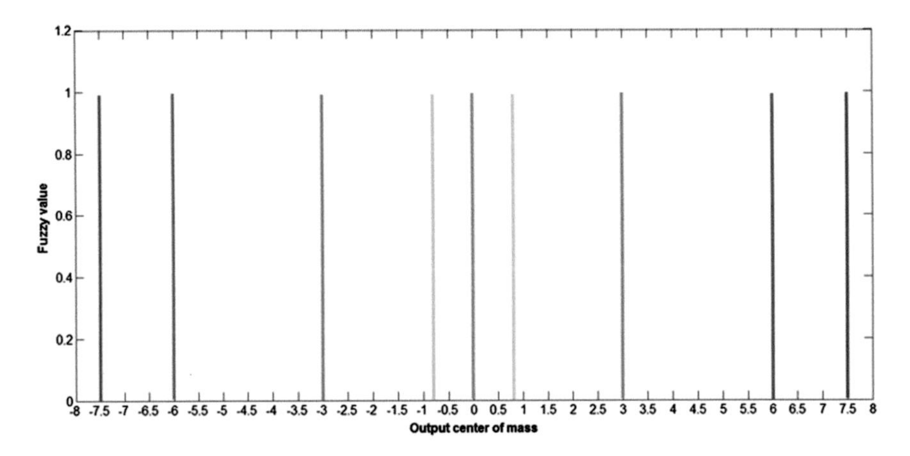

Fig. 1.38 Membership Functions for $M1$, $M2$, $M3$, and $M4$

Table 1.5 Inference rules

	Derivative					
Angle	NM	NS	Z	PS	PM	
	NL	PXL	PXL	PXL	PL	PL
	NM	PXL	PXL	PL	PM	PS
	NS	PXL	PL	PM	PS	NS
	Z	PM	PS	Z	NS	NM
	PS	PS	NS	NM	NL	NXL
	PM	NS	NM	NL	NXL	NXL
	PL	NL	NL	NXL	NXL	NXL

order to brake that rotation and try to compensate the angle (Motor 4 will receive the same value but negative to compensate the torque).

If the Quadrotor has a zero value angle (it is already aligned with the horizontal reference) and its derivative is zero (it is not rotating), then neither Motor 2 nor Motor 4 should receive a new value of PWM (see Fig. 1.40).

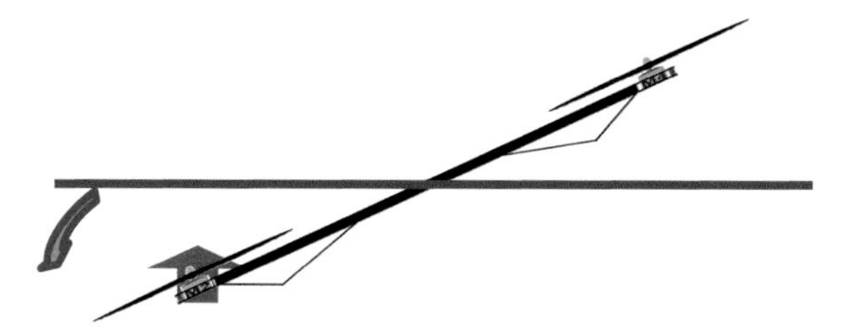

Fig. 1.39 Large angle and negative medium derivative

Fig. 1.40 Inference rule example 2

If all the conditions are evaluated, the inference table is obtained. Table 1.5 illustrates the rules implemented.

The surface generated with the controller designed is shown in Fig. 1.41. The membership functions proposed allow the system to move smoothly for reaching the position command signal, so the overshoot is reduced.

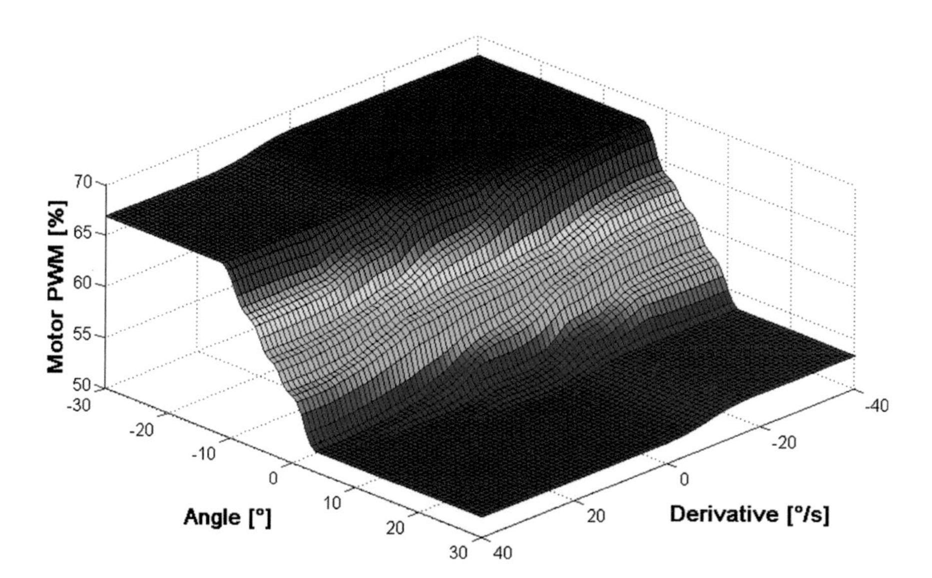

Fig. 1.41 Surface generated

The motors are affected by the corrective actions of the axis in which they are (x for motors 2 and 4 and y for motors 1 and 3) and by the corrective action in the z axis. Therefore, the final value for each motor will be given by the Eq. 1.56

$$\begin{aligned} M1 &= BV + X\text{corrector} + Z\text{corrector} \\ M2 &= BV + Y\text{corrector} - Z\text{corrector} \\ M3 &= BV - X\text{corrector} + Z\text{corrector} \\ M4 &= BV - Y\text{corrector} - Z\text{corrector} \end{aligned} \tag{1.56}$$

where BV is the base value of each motor.

1.17 Design of Fuzzy Logic Controller Tunned by an Anfis

If the surface generated by the fuzzy logic controller type 2 (see Fig. 1.41) is used for training the ANFIS, it is possible to reduce the number of membership functions and linguistic rules. The training method selected is the backpropagation. The ANFIS was trained with 3 membership functions in the angle and 3 membership functions in the derivative. It can be observed that the absolute minimum error reached is equal to 0.4577, which is a good result for flying the Quadrotor (see Fig. 1.42). If the number of membership functions is increased, the error decreased (see Fig. 1.43) but it is not the goal of this proposal. The target is to decrease the number of membership functions and rules for using lees computational resources. The final domains for the outputs and inputs trained by the ANFIS are presented in Figs. 1.44, 1.45, and 1.46.

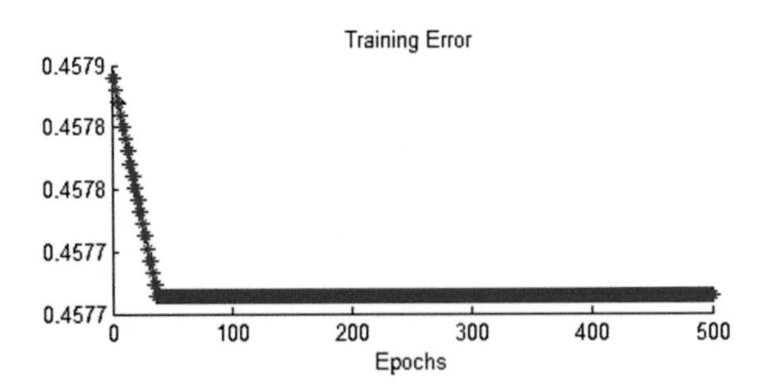

Fig. 1.42 Absolute minimum error reached with 3 membership functions for the angle and 3 membership functions for the derivative

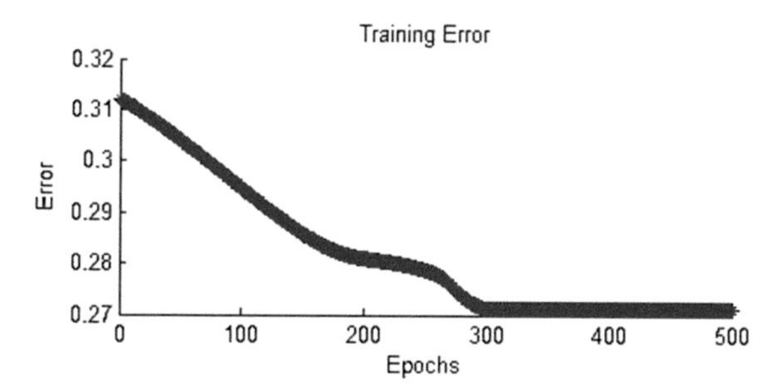

Fig. 1.43 Absolute minimum error reached with 7 membership functions for the angle and 5 membership functions for the derivative

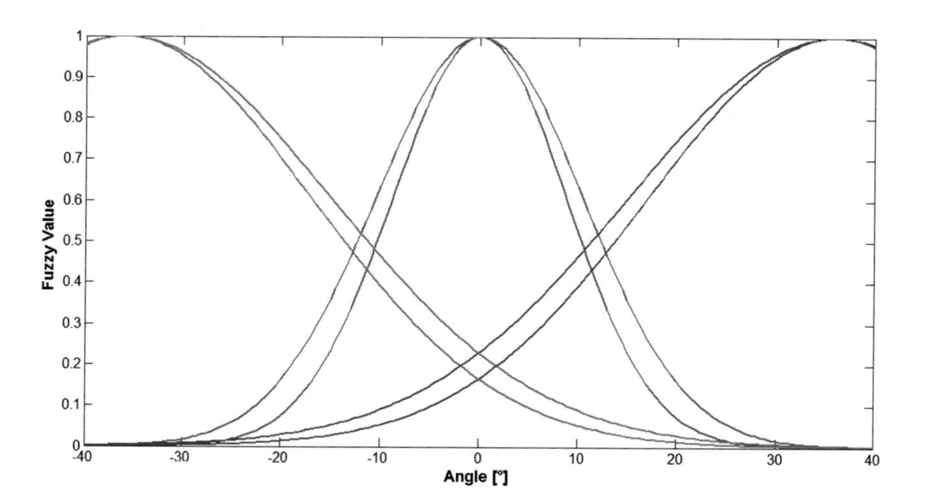

Fig. 1.44 Angle MFs obtained with ANFIS

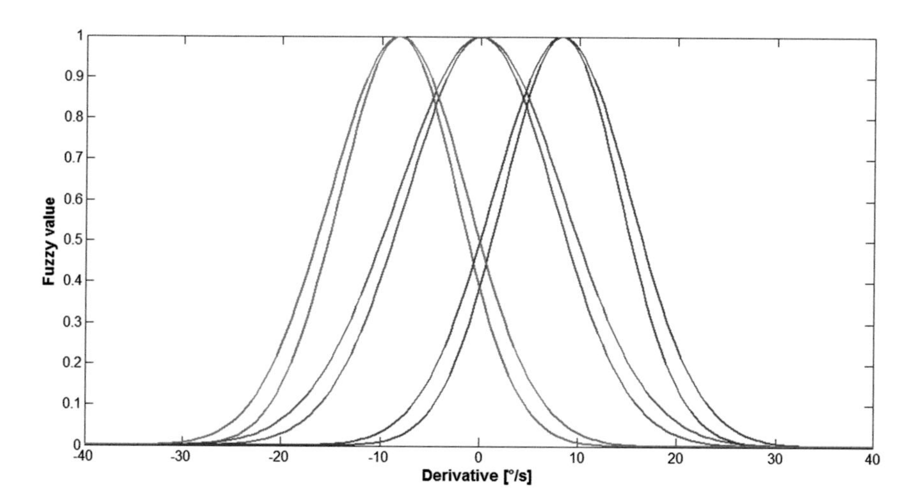

Fig. 1.45 Derivative MFs obtained with ANFIS

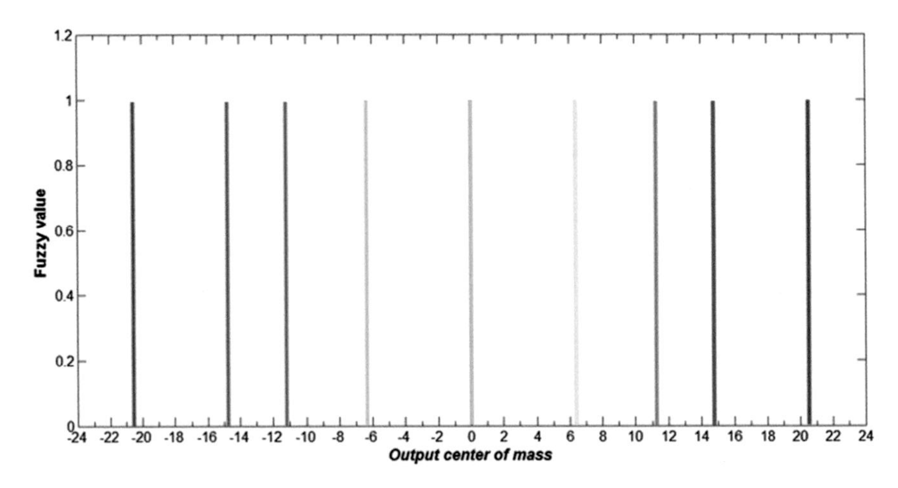

Fig. 1.46 Output MFs obtained with ANFIS

1.18 Experimental Results

Experimental results were created with different magnitude of noise added to the position sensor's signal (angle and derivative) and initial position equal to 30° is selected for all the tests. A LabVIEW frontal panel was developed for recording the information for each test; Fig. 1.47 shows the fontal panel for the Quadrotor.

When the noise was not presented, the error signal from the ANFIS tuned controller is better than the controller tuned by the expert. Figure 1.48 depicts the response for the controller tuned by the ANFIS (a) and tuned by the expert (b). It is observed that the fuzzy logic controller tuned by ANFIS the error is inside a band (±10°), but the system tuned by the expert some ripples occurred at sec. 22.

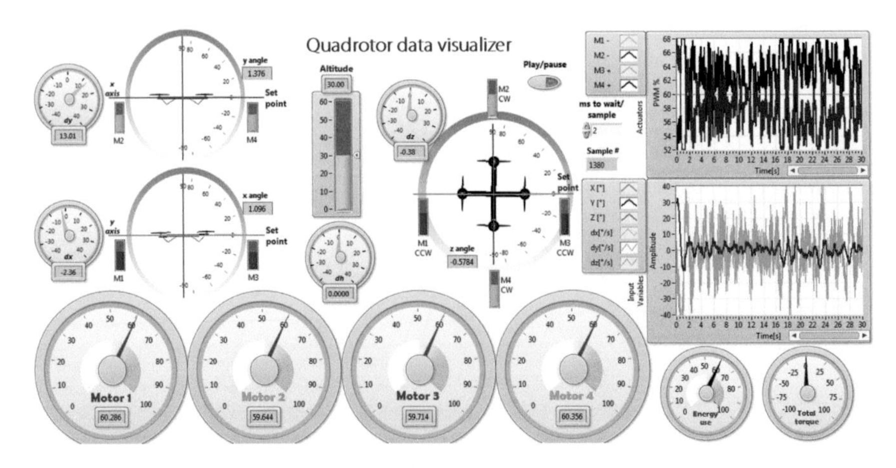

Fig. 1.47 LabVIEW frontal panel for the Quadrotor

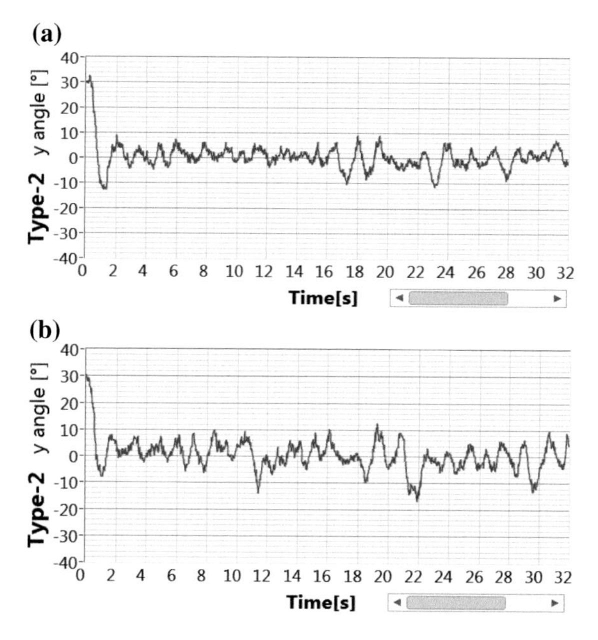

Fig. 1.48 No noise included in the system, controller tuned by ANFIS (**a**) and tuned by the expert (**b**)

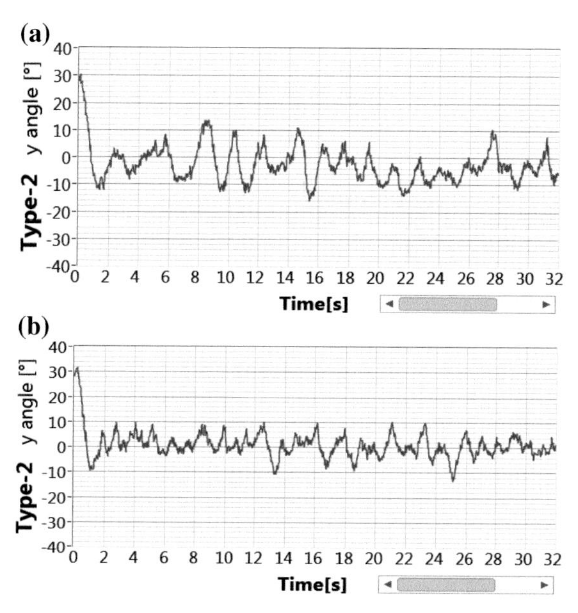

Fig. 1.49 Noise included in the system, controller tuned by ANFIS (**a**) and tuned by the expert (**b**) medium level of noise

When noise is added in both angle (±5°) and derivative (±10/s°), both controllers can deal with this noise level. Figure 1.49 shows how the response in the controller tuned by the ANFIS is degraded and the expert response can tolerate the noise in the band (±10°).

When noise is added in both angle (+/-25 degrees) and derivative (±50/s°), the fuzzy logic controller tuned with ANFIS gives a better response than the controller tuned by the expert (see Fig. 1.50) which has big ripples in the response.

It can be observed that the fuzzy logic controller tuned by ANFIS gives good response with no-noise and high level of noise but the response with medium level of noise is improved by the controller tuned by the expert because the number of membership functions generate smooth transitions. When more membership functions are implemented, the transition with medium level of noise is smoother than fuzzy logic controller tuned by ANFIS.

Table 1.6 shows how the processing time changes when a different fuzzy logic controller is implemented in the digital micro-controller. The BeagleBone Black (BBB) was the microprocessor selected. A list of some specifications of the board is presented below:

- Processor: AM3359 ARM Cortex-A8
- Speed Processor: 1 GHz
- Memory: 512 MB DDR3 (800 MHz x 16), 2 GB on-board storage using eMMC
- Digital pins: 65
- Analog pins: 7
- PWM pins: 8

In this case, it is clear that Type-2 optimized by ANFIS needs a lower processing time than Type-2 tuned by the expert. It is important to mention that fuzzy logic controller type 2 tuned by an expert is composed of 7 membership functions for the

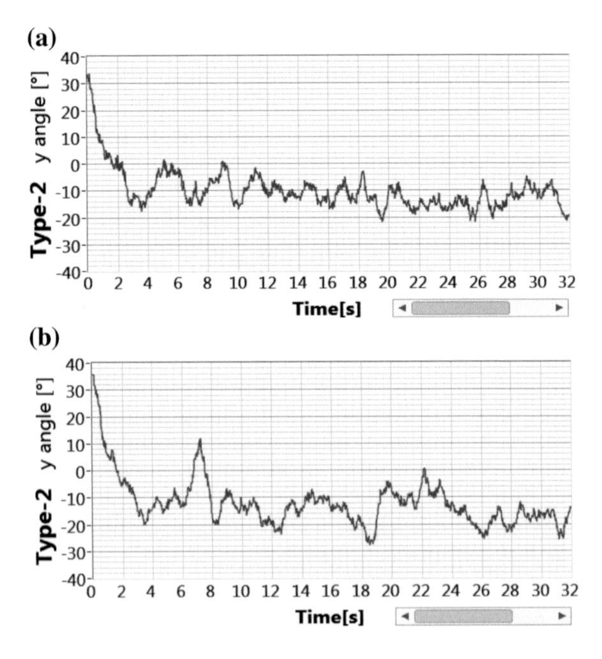

Fig. 1.50 No noise included in the system, controller tuned by ANFIS (**a**) and tuned by the expert (**b**) high level of noise

Table 1.6 Processing time for each fuzzy logic controller

	Type 2 7 and 5	ANFIS T2 3 and 3
	21.1	14.4
	21.8	14.7
	21.7	14.6
	20.9	14.7
	20.3	15.2
	21.4	13.6
	21.0	13.6
	20.0	15.4
	21.5	14.4
	21.6	14.7
Average	21.1	14.5

angle and 5 membership functions for the derivative, while the ANFIS tuned controller is composed only by 3 membership functions for the angle and 3 membership functions for the derivative.

References

1. K. Sudha, R. Vijaya Santhi, Robust decentralized load frequency control of interconnected power system with generation rate constraint using type-2 fuzzy approach. Int. J. Electr. Power Energy Syst. **33**, 699–707 (2011)
2. E.A. Jammeh, M. Fleury, C. Wagner, H. Hagras, M. Ghanbari, Interval type-2 fuzzy logic congestion control for video streaming across IP networks. IEEE Trans. Fuzzy Syst. **17**, 1123–1142 (2009)
3. C.-H. Lee, Y.-C. Lin, An adaptive type-2 fuzzy neural controller for nonlinear uncertain systems. Cont. Intell. Syst. **33**, 13–25 (2005)
4. M. Biglarbegian, W. Melek, J.M. Mendel, Design of novel interval type-2 fuzzy controllers for modular and reconfigurable robots: theory and experiments. IEEE Trans. Ind. Electron. **58**, 1371–1384 (2011)
5. G.O. Koca, Z.H. Akpolat, M. Özdemir, Type-2 fuzzy sliding mode control of a four-bar mechanism. Int. J. Model. Simul. **31**, 60 (2011)
6. K. Poornaselvan, T.G. Kumar, V.P. Vijayan, Agent based ground flight control using type-2 fuzzy logic and hybrid ant colony optimization to a dynamic environment, in *First International Conference on Emerging Trends in Engineering and Technology (ICETET'08)*, pp. 343–348 (2008)
7. M. Zaher, H. Hagras, Data generated type-2 fuzzy logic model for control of wind turbines, in *10th International Conference on Intelligent Systems Design and Applications (ISDA)*, pp. 80–85 (2010)
8. M. Galluzzo, B. Cosenza, Adaptive type-2 fuzzy logic control of a bioreactor. Chem. Eng. Sci. **65**, 4208–4221 (2010)
9. H. Chaoui, W. Gueaieb, Type-2 fuzzy logic control of a flexible-joint manipulator. J. Intell. Rob. Syst. **51**, 159–186 (2008)

10. D.V. Petrović, M. Tanasijević, V. Milić, N. Lilić, S. Stojadinović, I. Svrkota, Risk assessment model of mining equipment failure based on fuzzy logic. Expert Syst. Appl. **41**, 8157–8164 (2014)
11. M. Baldania, D. Sawant, A. Patki, Dynamic rule based approach to reduce power consumption of the fuzzy logic controller for embedded applications," in *First International Conference on Networks & Soft Computing (ICNSC)*, pp. 193–197 (2014)
12. V. Novak, V. Pavliska, R. Valasek, Specialized software for fuzzy natural logic and fuzzy transform applications, in *IEEE International Conference on Fuzzy Systems (FUZZ-IEEE)*, pp. 2337–2344 (2014)
13. W. Arbex, M. Martins, M.V. Silva, L.A. Carvalho, Fuzzy inference to decision support for SNP discovery, in *7th Iberian Conference on Information Systems and Technologies (CISTI)*, pp. 1–6 (2012)
14. W. Ai-Zhen, R.G. Feng, The Design of Neural Network Fuzzy Controller in Washing Machine, in *International Conference on Computing, Measurement, Control and Sensor Network (CMCSN)*, pp. 136–139 (2012)
15. Q. Liang, J.M. Mendel, Equalization of nonlinear time-varying channels using type-2 fuzzy adaptive filters. IEEE Trans. Fuzzy Syst. **8**, 551–563 (2000)
16. Q. Liang, J.M. Mendel, MPEG VBR video traffic modeling and classification using fuzzy technique. IEEE Trans. Fuzzy Syst. **9**, 183–193 (2001)
17. H.B. Mitchell, Pattern recognition using type-II fuzzy sets. Inf. Sci. **170**, 409–418 (2005)
18. P. Herman, G. Prasad, T. McGinnity, Investigation of the type-2 fuzzy logic approach to classification in an EEG-based brain-computer interface, in *27th Annual International Conference of the Engineering in Medicine and Biology Society (IEEE-EMBS 2005)*, pp. 5354–5357 (2006)
19. N.N. Karnik, J.M. Mendel, Applications of type-2 fuzzy logic systems to forecasting of time-series. Inf. Sci. **120**, 89–111 (1999)
20. T. Ozen, J.M. Garibaldi, Effect of type-2 fuzzy membership function shape on modelling variation in human decision making," in *IEEE International Conference on Fuzzy Systems*, pp. 971–976 (2004)
21. Q. Liang and L. Wang, "Sensed signal strength forecasting for wireless sensors using interval type-2 fuzzy logic system," in *The 14th IEEE International Conference on Fuzzy Systems (FUZZ'05)*, pp. 25–30 (2005)
22. O. Castillo and P. Melin, "Adaptive Noise Cancellation Using Type-2 Fuzzy Logic and Neural Networks," in *Type-2 Fuzzy Logic: Theory and Applications*, ed: Springer, 2008, pp. 213–223
23. C.-H. Lee, Y.-C. Lin, W.-Y. Lai, Systems identification using type-2 fuzzy neural network (type-2 FNN) systems, in *IEEE International Symposium on Computational Intelligence in Robotics and Automation*, pp. 1264–1269 (2003)
24. F.-H. Rhee, C. Hwang, An interval type-2 fuzzy perceptron, in *Proceedings of the 2002 IEEE International Conference on Fuzzy Systems (FUZZ-IEEE'02)*, pp. 1331–1335 (2002)
25. J.M. Keller, M.R. Gray, J.A. Givens, A fuzzy k-nearest neighbor algorithm. Syst. Man Cybern. IEEE Trans. (4), 580–585 (1985)
26. J. Agero, A. Vargas, Using type-2 fuzzy logic systems to infer the operative configuration of distribution networks," in *Proceedings IEEE Power Engineering Society General Meeting*, pp. 2379–2386 (2005)
27. N. Sulaiman, Z.A. Obaid, M. Marhaban, M. Hamidon, FPGA-based fuzzy logic: design and applications–a review. IACSIT Int. J. Eng. Technol. **1**, 491–503 (2009)
28. G. Muscato, Fuzzy control of an underactuated robot with a fuzzy microcontroller. Microprocess. Microsyst. **23**, 385–391 (1999)
29. L. Faravelli, R. Rossi, G. Torelli, Numerical testing of a programmable microcontroller with fuzzy and adaptive features. Simul. Model. Pract. Theory **11**, 421–431 (2003)
30. A.M. El-Nagar, M. El-Bardini, Practical Implementation for the interval type-2 fuzzy PID controller using a low cost microcontroller. Ain Shams Eng. J. **5**, 475–487 (2014)

31. G. Bal, E. Bekiroğlu, Ş. Demirbaş, I. Colak, Fuzzy logic based DSP controlled servo position control for ultrasonic motor. Energy Convers. Manag. **45**, 3139–3153 (2004)
32. Y.L. Goh, A.K. Ramasamy, F.H. Nagi, A.A. Zainul Abidin, DSP based fuzzy and conventional overcurrent relay controller comparisons. Microelectron. Reliab. **53**, 1029–1035 (2013)
33. Y.L. Goh, A.K. Ramasamy, F.H. Nagi, A.A.Z. Abidin, DSP based overcurrent relay using fuzzy bang–bang controller. Microelectron. Reliab. **51**, 2366–2373 (2011)
34. P.-H. Chou, C.-S. Chen, F.-J. Lin, DSP-based synchronous control of dual linear motors via Sugeno type fuzzy neural network compensator. J. Franklin Inst. **349**, 792–812 (2012)
35. M.A. Melgarejo R, C.A. Peña-Reyes, Hardware architecture and FPGA implementation of a type-2 fuzzy system, in *Proceedings of the 14th ACM Great Lakes Symposium on VLSI*, pp. 458–461 (2004)
36. E. Ramadan, M. El-bardini, M. Fkirin, Design and FPGA-implementation of an improved adaptive fuzzy logic controller for DC motor speed control. Ain Shams Eng. J. (2014)
37. A. Messai, A. Mellit, A. Guessoum, S. Kalogirou, Maximum power point tracking using a GA optimized fuzzy logic controller and its FPGA implementation. Sol. Energy **85**, 265–277 (2011)
38. M.P. Soares dos Santos, J. Ferreira, Novel intelligent real-time position tracking system using FPGA and fuzzy logic. ISA Trans. **53**, 402–414 (2014)
39. D.M. Munoz, C.H. Llanos, M. Ayala-Rincon, R.H. van Els, Distributed approach to group control of elevator systems using fuzzy logic and FPGA implementation of dispatching algorithms. Eng. Appl. Artif. Intell. **21**, 1309–1320 (2008)
40. L.A. Zadeh, Fuzzy sets. Inf. Control **8**, 338–353 (1965)
41. L.A. Zadeh, Fuzzy logic = computing with words. IEEE Trans. Fuzzy Syst. **4**, 103–111 (1996)
42. B.K. Bose, *Power electronics and variable frequency drives: technology and applications* (IEEE Press, New Jersey, 2010)
43. M. Zamani, H. Nejati, A.T. Jahromi, A. Partovi, S.H. Nobari, G.N. Shirazi, Toolbox for Interval Type-2 Fuzzy Logic Systems, in *11th Joint International Conference on Information Sciences* (2008)
44. E.H. Mamdani, Application of fuzzy algorithms for control of simple dynamic plant," in *Proceedings of the Institution of Electrical Engineers*, pp. 1585–1588 (1974)
45. D. Wu, W.W. Tan, A simplified architecture for type-2 FLSs and its application to nonlinear control, in *IEEE Conference on Cybernetics and Intelligent Systems*, pp. 485–490 (2004)
46. M. Mizumoto, K. Tanaka, Some properties of fuzzy sets of type 2. Inf. Control **31**, 312–340 (1976)
47. J.M. Mendel, H. Hagras, W.W. Tan, W. Melek, H. Ying, *Introduction to Type-2 Fuzzy Logic Control: Theroy and Applications* (Wiley, New York, 2014)
48. T. Takagi, M. Sugeno, Fuzzy identification of systems and its applications to modeling and control. Systems, IEEE Trans. Man Cybern. 116–132 (1985)
49. M. Sugeno, G. Kang, Structure identification of fuzzy model. Fuzzy Sets Syst. **28**, 15–33 (1988)
50. F. Liu, An efficient centroid type-reduction strategy for general type-2 fuzzy logic system. Inf. Sci. **178**, 2224–2236 (2008)
51. L.A. Zadeh, The concept of a linguistic variable and its application to approximate reasoning —I. Inf. Sci. **8**, 199–249 (1975)
52. N.N. Karnik, J.M. Mendel, Centroid of a type-2 fuzzy set. Inf. Sci. **132**, 195–220 (2001)
53. P. Ponce, *Inteligencia artificial con aplicaciones a la ingeniería*: Alfaomega (2011)
54. J.M. Mendel, *Uncertain Rule-Based Fuzzy Logic System: Introduction and New Directions* (2001)
55. L.A. Zadeh, Outline of a new approach to the analysis of complex systems and decision processes, IEEE Trans. Syst. Man Cybern. 28-44 (1973)
56. N.N. Karnik, J.M. Mendel, Q. Liang, Type-2 fuzzy logic systems. IEEE Trans. Fuzzy Syst. **7**, 643–658 (1999)

57. M.B. Gorzałczany, A method of inference in approximate reasoning based on interval-valued fuzzy sets. Fuzzy Sets Syst. **21**, 1–17 (1987)
58. D.G. Schwartz, The case for an interval-based representation of linguistic truth. Fuzzy Sets Syst. **17**, 153–165 (1985)
59. I. Türkşen, Fuzzy normal forms. Fuzzy Sets Syst. **69**, 319–346 (1995)
60. G.J. Klir, T.A. Folger, *Fuzzy Sets, Uncertainty, and Information* (1988)
61. Q. Liang, J.M. Mendel, Interval type-2 fuzzy logic systems: theory and design. IEEE Trans. Fuzzy Syst. **8**, 535–550 (2000)
62. H. Wu, J.M. Mendel, Uncertainty bounds and their use in the design of interval type-2 fuzzy logic systems. IEEE Trans. Fuzzy Syst. **10**, 622–639 (2002)
63. J.M. Mendel, On a 50% savings in the computation of the centroid of a< i> symmetrical</i> interval type-2 fuzzy set. Inf. Sci. **172**, 417–430 (2005)
64. S. Coupland, R. John, A fast geometric method for defuzzification of type-2 fuzzy sets. IEEE Trans. Fuzzy Syst. **16**, 929–941 (2008)
65. D. Wu, J.M. Mendel, Enhanced karnik–mendel algorithms. IEEE Trans. Fuzzy Syst. **17**, 923–934 (2009)
66. H.-J. Wu, Y.-L. Su, S.-J. Lee, A fast method for computing the centroid of a type-2 fuzzy set. IEEE Trans. Syst. Man Cybern. Part B: Cybern. **42**, 764–777 (2012)
67. J.M. Mendel, R.B. John, Type-2 fuzzy sets made simple. IEEE Trans. Fuzzy Syst. **10**, 117–127 (2002)
68. N.N. Karnik, J.M. Mendel, Introduction to type-2 fuzzy logic systems," in *The 1998 IEEE International Conference on Fuzzy Systems Proceedings, IEEE World Congress on Computational Intelligence*, pp. 915–920 (1998)
69. N.N. Karnik, J.M. Mendel, Operations on type-2 fuzzy sets. Fuzzy Sets Syst. **122**, 327–348 (2001)
70. C. Coza, C.J.B. Macnab, A new robust adaptive-fuzzy control method applied to quadrotor helicopter stabilization, in *Fuzzy Information Processing Society, 2006. NAFIPS 2006. Annual meeting of the North American*, pp. 454–458. IEEE (2006, June)
71. M.L. Santos, Intelligent fuzzy controller of a quadrotor. *IEEE*, pp. 141–146 (2010)
72. A.R. Kirli, Self tuning fuzzy PD application on TI TMS320F28335 for an experimental stationary Quadrotor. Yildiz Tech. Univ (2010)
73. S.S. Sheikhpour, a model-based fuzzy controller using the parallel distributed compensation method for quadrotorattitude stabilization. *Sharif University of Technology. Anticlockwise Roll, D: Clockwise Roll, E: Anticlockwise Pitch, F: Clockwise Pitch* (2013)
74. I.K. Ilhan, Type-2 Fuzzy based quadrotorcontrol approach, in *9th Asian Control Conference (ASCC)* (IEEE, Istanbul, 2013), pp. 1–6
75. P. Bhatkhande, T.C. Havens, Real time fuzzy controller for quadrotor stability control, in *Fuzzy Systems (FUZZ-IEEE), 2014 IEEE International Conference on*, pp. 913–919. IEEE (2014, July)
76. G.D. Qingji, Research of stability augmentation hybrid controller for quadrotor UAV, in *26th Chinese Control and Decision Conference (CCDC)* (IEEE, Tianjin, 2014), pp. 5224–5229
77. J.M. Brito, *Quadrotor Prototype. Dissertação para obtenção do Grau de M[estre.* Lisboa: Universidad Técnica de Lisboa (2009)
78. G.H. Hoffmann, QuadrotorHelicopter Flight Dynamics and Control:Theory and Experiment, in *AIAA Guidance, Navigation and Control Conference and Exhibit* (American Institute of Aeronautics and Astronautics, South Carolina, 2007)
79. J.M. Mendel, R. John, F. Liu, Interval type-2 fuzzy logic systems made simple. Fuzzy Syst. IEEE Trans. **14**(6), 808–821 (2006)
80. J.M. Mendel, Type-2 fuzzy sets and systems: an overview. Comput. Intell. Mag. IEEE**2**(1), 20–29 (2007)
81. J.-S.R. Jang, ANFIS: Adaptive-Network-based Fuzzy Inference Systems. IEEE Trans. Syst. Man Cybern. 665–685 (1993)

Chapter 2
LabVIEW™ FPGA

2.1 Field-Programmable Gate Array (FPGA)

An field-programmable gate array (FPGA) [1] is a device that contains a matrix of reconfigurable gate array logic circuitry. When an FPGA is configured, the internal circuitry is connected in a way that creates a hardware implementation of the software application. Unlike processors, FPGAs use dedicated hardware for processing logic and do not have an operating system. FPGAs are truly parallel in nature, so different processing operations do not have to compete for the same resources. As a result, the performance of one part of the application is not affected when additional processing is added. Also, multiple control loops can run on a single FPGA device at different rates. FPGA-based control systems can enforce critical interlock logic and can be designed to prevent I/O forcing by an operator. However, unlike hard-wired printed circuit board (PCB) designs which have fixed hardware resources, FPGA-based systems can literally rewire their internal circuitry to allow reconfiguration after the control system is deployed to the field. FPGA devices deliver the performance and reliability of dedicated hardware circuitry.

A single FPGA can replace thousands of discrete components by incorporating millions of logic gates in a single integrated circuit (IC) chip. The internal resources of an FPGA chip consist of a matrix of configurable logic blocks (CLBs) surrounded by a periphery of I/O blocks. Signals are routed within the FPGA matrix by programmable interconnect switches and wire routes (see Fig. 2.1).

FPGA technology provides the reliability of dedicated hardware circuitry, true parallel execution, and lightning fast closed-loop control performance. This application note provides answers to frequently asked questions (FAQs) regarding the use of reconfigurable FPGA-based hardware targets for closed-loop control applications. A compactRIO is shown in Fig. 2.2.

© Springer International Publishing Switzerland 2016

P. Ponce-Cruz et al., *Fuzzy Logic Type 1 and Type 2 Based on LabVIEW™ FPGA*, Studies in Fuzziness and Soft Computing,
DOI 10.1007/978-3-319-26656-5_2

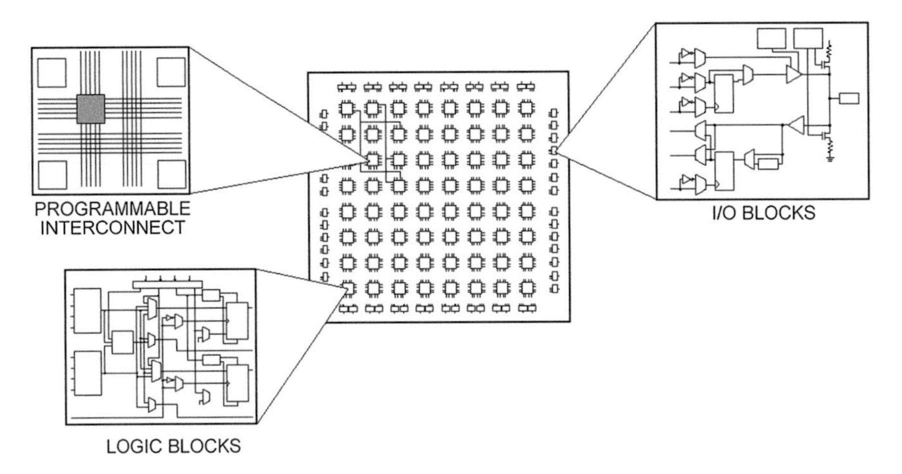

Fig. 2.1 Looking inside an FPGA chip

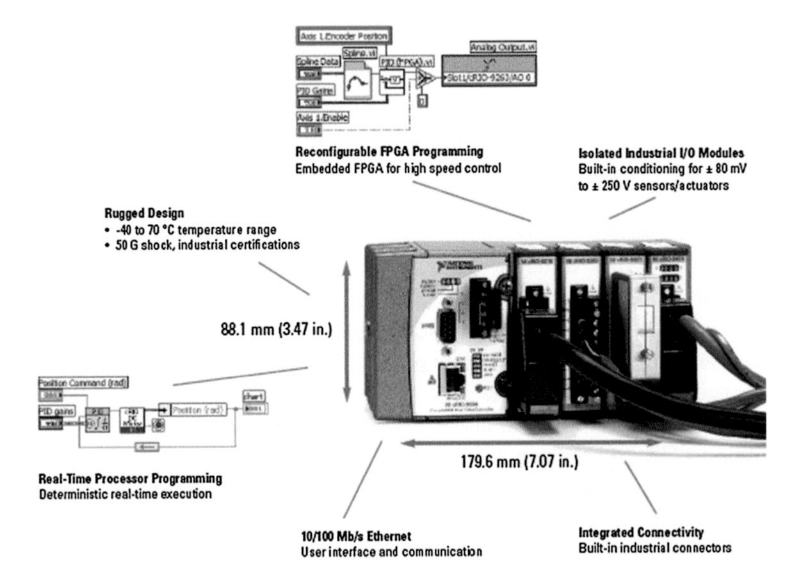

Fig. 2.2 NI CompactRIO is a small, rugged FPGA-based control system

2.1.1 How Do FPGA-Based Control Systems Compare to Processor-Based Systems?

Like processor-based control systems, FPGAs have been used to implement all types of industrial control systems (see Fig. 2.3), including analog process control, discrete logic, and batch or state machine-based control systems. However, FPGA-based control systems differ from processor-based systems in significant ways.

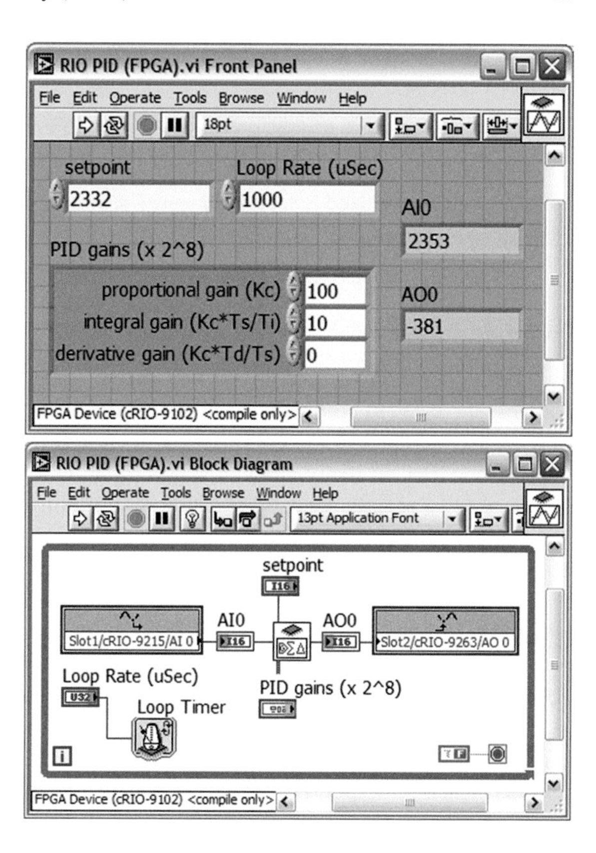

Fig. 2.3 Performing PID control in LabVIEW FPGA

When you compile [2] your control application for an FPGA device, the result is a highly optimized silicon implementation that provides true parallel processing with the performance and reliability benefits of dedicated hardware circuitry. Since there is no operating system on the FPGA chip, the code is implemented in a way that ensures maximum performance and reliability.

In addition to offering high reliability, FPGA devices can perform deterministic closed-loop control at extremely fast loop rates. In most FPGA-based control applications, speed is limited by the sensors, actuators, and I/O modules rather than the processing performance of the FPGA. For example, the proportional-integral-derivative (PID) control algorithm that is included with the LabVIEW FPGA Module executes in just 300 ns (0.000000300 s). PID control is commonly used for regulating analog process values such as pressure, temperature, force, displacement, fluid flow, or electrical current.

FPGA-based control systems offer deterministic closed control performance at rates exceeding 1 MHz. In fact, many algorithms can be executed in a single cycle of the FPGA clock (40 MHz). Processing is done in parallel, so multi-rate control

Processor-based control		FPGA-based control
Performance limited to 1 kHz	→	Closed loop performance beyond 1 MHz
Serial execution, single rate control	→	Parallel execution, multi-rate control
Performance slows as app. grows	→	No slow down as application grows
Operating system runs control logic	→	Control logic in dedicated hardware
I/O modules have fixed functionality	→	I/O functionality is reconfigurable
Custom circuitry requires board layout	→	Software defined gate array
Separate motion control system	→	Motion integrated with other control logic

Fig. 2.4 Processor-based control (*left*) compared to FPGA-based control (*right*)

systems are easy to implement. Since control logic runs in dedicated hardware subsystems on the FPGA, applications do not slow down when additional processing is added. In many cases, a software-defined gate array in FPGA hardware can be used to replace a costly and time-consuming custom PCB layout.

FPGAs can digitally process signals at very high speeds and are often used to reconfigure the I/O module functionality. For example, a digital input module can be used to simply read the true/false state of each digital line. Alternately, the same FPGA can be reconfigured to perform processing on the digital signals and measure pulse width, perform digital filtering, or even measure position and velocity from a quadrature encoder sensor.

FPGA-based systems often incorporate motion control and motor drive commutation in a single FPGA-based control application. By contrast, processor-based systems typically offload the motor drive commutation to separate hardware since motor current or torque control requires fast loop rates (commonly 20 kHz) and precise timing of the gate drive commutation signals. A comparison between processors controller and LabVIEW FPGA is presented in Fig. 2.4.

2.1.2 How Do I Program My Control Application Using the LabVIEW FPGA Module?

The LabVIEW FPGA Module enables you to use high-level graphical dataflow programming to create a highly optimized gate array implementation of your analog or digital control logic. You can use normal LabVIEW programming techniques to develop your FPGA application. When you target FPGA hardware such as a CompactRIO chassis or R Series intelligent data acquisition (DAQ) device, the LabVIEW programming palette is simplified to contain only the functions that are designed to work on FPGAs [3]. The primary programming difference compared to traditional LabVIEW is that FPGA devices use integer math rather than floating-point math. Also, there is no notion of multithreading or priorities since each loop executes in independent dedicated hardware and does not have share resources—in effect, each loop executes in parallel at “time critical” priority.

The LabVIEW FPGA palette contains extensive intellectual property (IP) libraries [4]. Table 2.1 shows a list of some of the key function blocks for developing FPGA-based control systems. For more details, see the LabVIEW FPGA Module user manual in NI website.

Table 2.1 Lists of some of the key function blocks for developing FPGA-based control systems

Category	Key functions for control	Common control applications
Programming structures	For Loop, While Loop, Case Structure, Feedback Node, Sequence Structure, Single Cycle Timed Loop, Shift Register, HDL Interface Node	Analog process control loops, state machines, batch control, sequential function charts, event response, repeated execution, signal latching, subroutines, sequencing, system state control (power up, shut down, watchdog, fault, …)
Input/Output	Analog Input, Analog Output, Digital Input, Digital Output, Digital Port Input, Digital Port Output	Interfacing to digital I/O, voltage, current, temperature, load, pressure, strain, relay, 4–20 mA, H-bridge, CAN communication, wireless networking, and other signals
Analog control	Discrete PID, Discrete Control Filter, Discrete Delay, Discrete Normalized Integrator, Initial Condition, Unit Delay, Zero-Order Hold, Backlash, Dead Zone, Friction, Memory Element, Quantizer, Rate Limiter, Relay, Saturation, Switch, Trigger, Linear Interpolation, Sine Generator, Look-Up Table 1D	Analog control algorithms, filtering of noisy signals, limiting input/output signals, scaling nonlinear sensor signals to engineering unit proportional values, function generation, sine, cosine, log, exponential, gain scheduling, ramp/soak
Discrete logic	And, And Array Elements, Boolean Array To Number, Boolean To (0,1), Compound Arithmetic, Exclusive Or, Implies, Not, Not And, Not Exclusive Or, Not Or, Number To Boolean Array, Or, Or Array Elements, Boolean Crossing	Digital control, digital logic, Boolean logic, relay ladder logic, sequence of events, state transitions, control of 2-state and 3-state discrete devices, edge detection
Comparison functions	Equal?, Equal To 0?, Greater?, Greater Or Equal?, Greater Or Equal To 0?, Greater Than 0?, Less?, Less Or Equal?, Less Or Equal To 0?, Less Than 0?, Not Equal?, Not Equal To 0?, Select, Max and Min, In Range and Coerce, Zero Crossing	Alarming, triggering, event detection, peak detection, signal comparison, thresholding, change of state detection, signal selection (high, min, max), limit testing, selector/multiplexer, heating/cooling split range control

(continued)

Table 2.1 (continued)

Category	Key functions for control	Common control applications
Math	Absolute Value, Add, Compound Arithmetic, Decrement, Increment, Multiply, Negate, Quotient and Remainder, Scale By Power Of 2, Sign, Subtract, Saturation Add, Saturation Multiply, Saturation Subtract, Join Numbers, Logical Shift, Rotate, Rotate Left With Carry, Rotate Right With Carry, Split Number, Swap Bytes, Swap Words	Analog signal manipulation, summing, counter/timers, rate of change detection, electronic gearing/camming, accumulator, averaging, totalizer, digital signal processing
Data transfer, timing, triggering and synchronization	Global Variable, Local Variable, FIFO Read, FIFO Write, Memory Read, Memory Write, Interrupt, Loop Timer, Tick Count, Wait, Generate Occurrence, Set Occurrence, Wait On Occurrence, First Call?	Watchdogs, timers, accumulators, pulse width measurement/generation, timer on/off delay
NI SoftMotion module	Motion ControlLoop PID (32-bit), Spline Engine (Interpolation)	Multiaxis coordinated motion control, trajectory generation, straight line moves, jogging, arc move, contouring, interpolation
Digital filter design toolkit	Filter Design, Fixed-Point Tools, Code Generation	Digital filter design, convert floating-point to fixed-point, generate LabVIEW FPGA code

2.1.3 How Does the LabVIEW Compiler Translate My Graphical Code into FPGA Circuitry?

The LabVIEW FPGA module compiles your LabVIEW application to FPGA hardware using an automatic multistep process [2]. Behind the scenes, your graphical code is translated to text-based VHDL code. Then industry standard Xilinx ISE compiler tools are invoked and the VHDL code is optimized, reduced, and synthesized into a hardware circuit realization of your LabVIEW design. This process also applies timing constraints to the design and tries to achieve an efficient use of FPGA resources (sometimes called "fabric").

A great deal of optimization is performed during the FPGA compilation process to reduce digital logic and create an optimal implementation of the LabVIEW application (see Fig. 2.5). Then the design is synthesized into a highly optimized silicon implementation that provides true parallel processing capabilities with the performance and reliability of dedicated hardware [5].

The end result is a bit stream file that contains the gate array configuration information. When you run the application, the bit stream is loaded into the FPGA

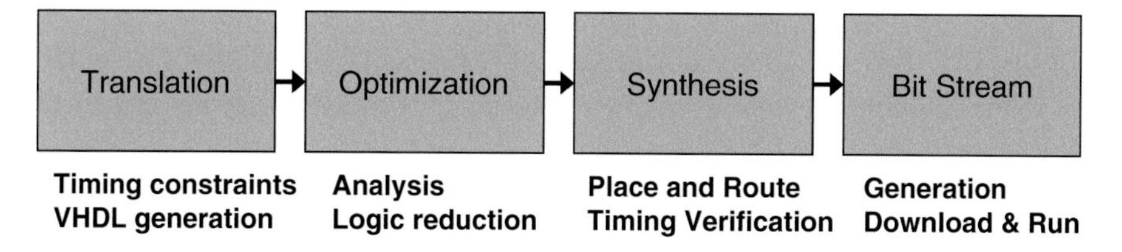

Fig. 2.5 LabVIEW FPGA compilation process

chip and used to reconfigure the gate array logic. The bit stream can also be loaded into nonvolatile flash memory and loaded instantaneously when power is applied to the target. There is no operating system on the FPGA chip; however, execution can be started and stopped using enable-chain logic that is built into the FPGA application.

2.1.4 *FPGAs Are Fast, but How Do Faster Loop Rates Improve Control System Performance?*

In general, the speed of the control system impacts its performance, stability, robustness, and disturbance rejection characteristics (see Fig. 2.6). Faster control systems are typically more stable, easier to tune, and less susceptible to changing conditions and disturbances.

To provide stable and robust control, a control system must be able to measure the process variable and set an actuator output command within a fixed period of time. Systems (plants) that can change quickly require fast control systems to guarantee reliable performance within acceptable limits. As a rule, the control loop rate should be at least ten times faster than the time constant of the system (plant). The time constant is a measure of the speed of the system.

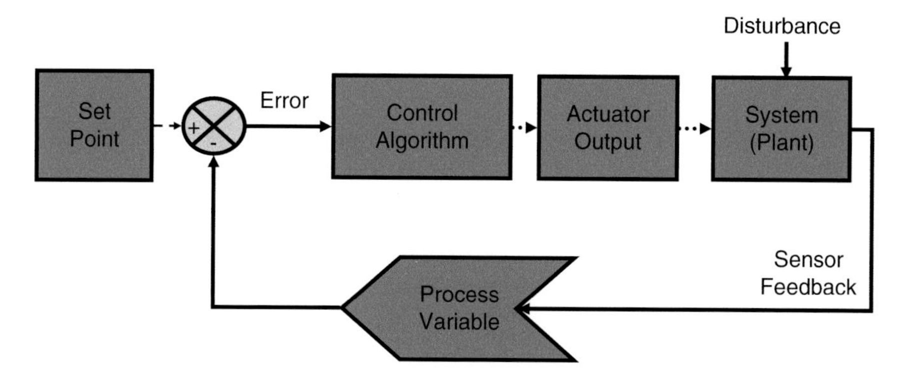

Fig. 2.6 Typical closed-loop control system

For example, the current in a DC motor may change as fast as 1 A per millisecond in response to a 24 V output from an H-bridge driver. To precisely control the motor current, the control system must sample the current quickly and make frequent adjustments to the actuator output.

2.1.5 *What FPGA Hardware Targets Are Available from NI?*

The CompactRIO reconfigurable embedded system (see Fig. 2.7) is a small modular system for industrial applications that require the highest level of ruggedness and reliability. CompactRIO is designed for harsh environments and offers a wide temperature range, high shock and vibration ratings, and an array of industrial certifications and ratings. CompactRIO is rated for marine environments, Class I, Division 2 rating for hazardous locations, and offers up to 2300 V of isolation. Like all FPGA targets from NI, CompactRIO uses the C Series industrial I/O modules for low-cost connectivity directly to industrial control sensors and actuators. In addition, there are many third-party vendors around the world that offer C Series I/O and communication modules.

The NI R Series intelligent DAQ devices are plug-in boards for PCI and PXI/CompactPCI buses with onboard FPGA hardware for user-defined signal processing and control. Up to 8 analog inputs, 8 analog outputs and 160 digital I/O channels are built into the intelligent DAQ devices. You can also connect an expansion chassis to any digital port and add C series industrial I/O modules. The NI intelligent DAQ devices enable you to define your own hardware functionality and offer limitless possibilities for timing, triggering, synchronization, digital signal processing, and control.

The PXI R Series intelligent DAQ system offers FPGA performance and reliability in the industry standard PXI form factor (see Fig. 2.8). In addition to the intelligent DAQ devices from NI, hundreds of non-reconfigurable plug-in boards are available from NI and other vendors around the world. The PXI system can be booted into Windows or the LabVIEW Real-Time operating system. C Series I/O modules provide signal conditioning and combine instrumentation grade accuracy

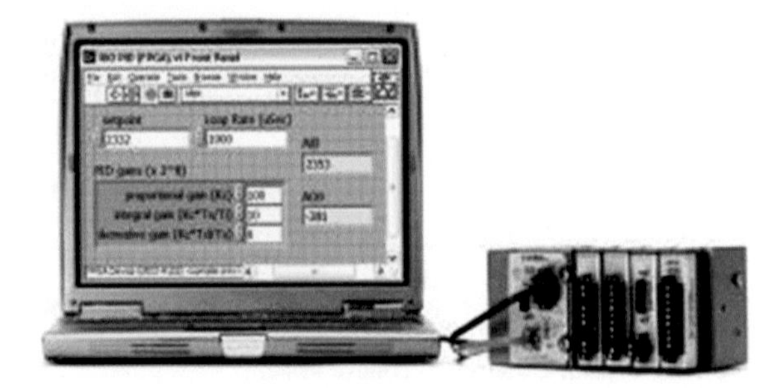

Fig. 2.7 NI CompactRIO reconfigurable embedded system

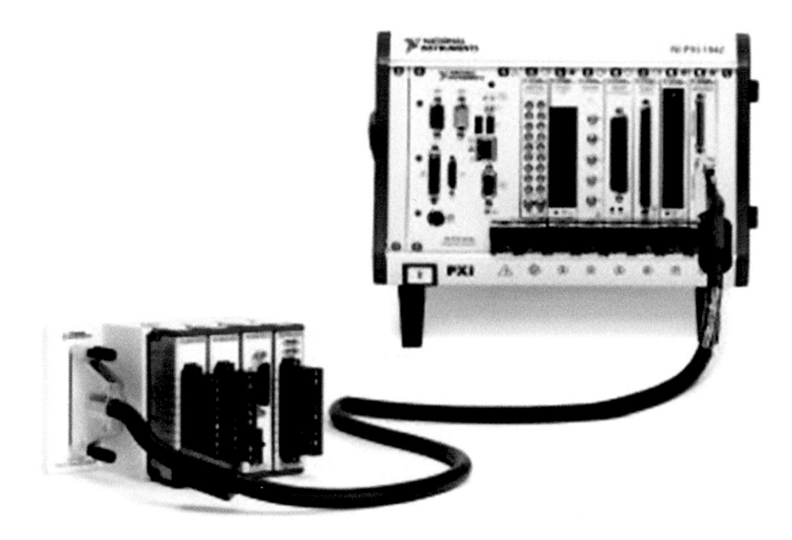

Fig. 2.8 PXI R series intelligent DAQ system

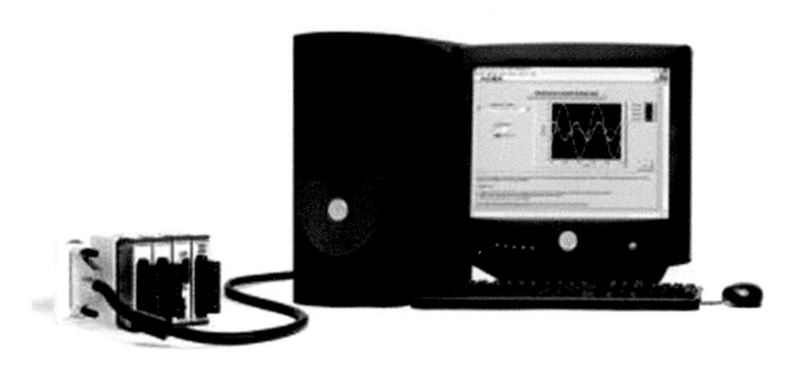

Fig. 2.9 PCI R series intelligent DAQ system

with industrial features such as isolation or high current drive capability. The R Series Expansion Chassis is used to connect C Series modules to intelligent DAQ devices. For more information, see the online application notes explaining the R Series Intelligent DAQ devices.

The PCI R Series Intelligent DAQ System enables you to add FPGA-based control capabilities to any desktop, industrial PC, or single-board computer (SBC) containing a PCI slot. Like all NI FPGA targets, the intelligent DAQ devices can load their bit stream instantly at power up from nonvolatile flash storage located on the plug-in board (see Fig. 2.9).

The National Instruments Compact Vision System (see Fig. 2.10) is a rugged standalone platform for industrial machine vision and I/O applications such as robotics, automated test, and automated inspection. All Compact Vision Systems

Fig. 2.10 NI compact vision system

contain a user-programmable FPGA for implementing custom triggers, counters, pulse width modulation (PWM), motion, and other digital control operations. NI Compact Vision systems use IEEE 1394 (FireWire) technology for interfacing to more than 300 compatible cameras.

2.1.6 *What Closed-Loop Control Performance Can I Achieve?*

In most cases, the computational performance of the FPGA is so fast that the control loop rate is limited only by the sensors, actuators, and I/O modules (see Fig. 2.11). This is a stark contrast to traditional control systems, where the processing performance was typically the limiting factor.

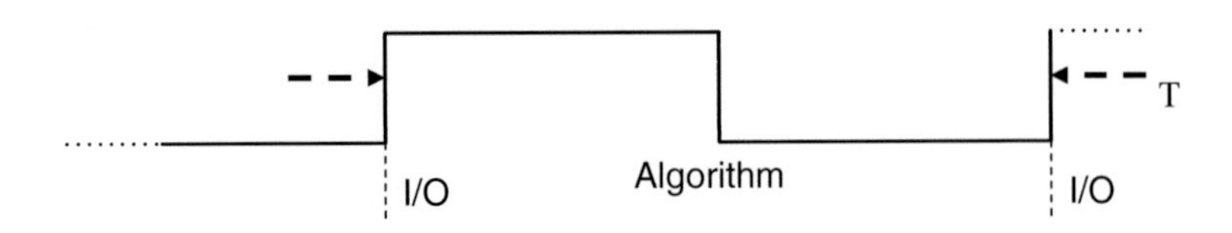

Fig. 2.11 The loop cycle time (*T*) is the time taken to execute one cycle of a control loop

For example, using R Series intelligent DAQ devices, the input/output and control logic calculations for discrete control applications can all be implemented at a 20 MHz control loop rate using the 5 V TTL digital I/O lines on the boards. These digital lines can be accessed from within a LabVIEW single-cycle timed loop (SCTL) executing at a 25 ns rate. Significant amounts of control logic can usually be included in a SCTL.

For 24 V discrete logic control applications using high current C Series digital I/O modules, the loop rate is limited to the update rate of the modules. For example, the NI 9423 digital input and NI 9474 digital output modules both have 1 μs update rates, resulting in a maximum 24 V discrete control performance of 500 kHz.

In analog process control applications, the control loop rate is also limited by the update rate of the I/O modules. The NI 9215 analog input and NI 9263 analog output modules offer 16-bit resolution and simultaneous sampling capabilities at 10 μs update rates. This results in a closed-loop analog process control performance of 50 kHz.

2.1.7 *How Much Jitter Can I Expect in My FPGA-Based Control Loops?*

A common gauge of control system performance and robustness is jitter (see Fig. 2.12), which is a measure of the variation of the actual loop cycle time from the desired loop cycle time. In general, purpose operating systems such as Windows, the jitter is unbounded so closed-loop control system stability cannot be guaranteed. Processor-based control systems with real-time operating systems are commonly able to guarantee control loop jitter of less than 100 μs. In FPGA-based applications, the control loop does not need to share hardware resources with other tasks and control loops can be precisely timed using the FPGA clock. The jitter for FPGA-based control loops depends on the accuracy of the FPGA clock source. In

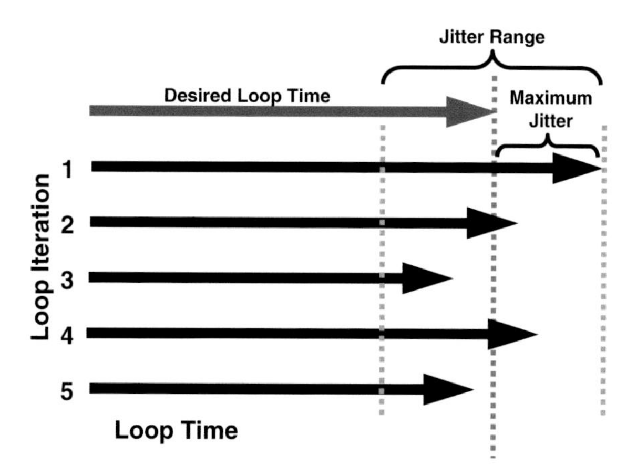

Fig. 2.12 To guarantee stability, control loop jitter must be bounded

the case of the CompactRIO cRIO-910x reconfigurable chassis, the FPGA clock jitter is only 250 ps (0.000000000250 s) when using a 40 MHz FPGA clock rate.

2.1.8 Creating a New LabVIEW Real-Time Project and Adding I/O

Now it is presented how the real-time system is included in a project [6]. This step allows to configure inputs and outputs (I/O) in real time.

1. Launch **NI LabVIEW** by clicking on the desktop icon. Then click on the **Real-Time Project** (see Fig. 2.13) link to start a new LabVIEW project for your NI CompactRIO system.
   LabVIEW 8.20 has a Real-Time Project Wizard that makes creating and configuring real-time applications easy. To help you get started, the wizard enables you to choose an appropriate programming architecture and automatically generates a software template application.

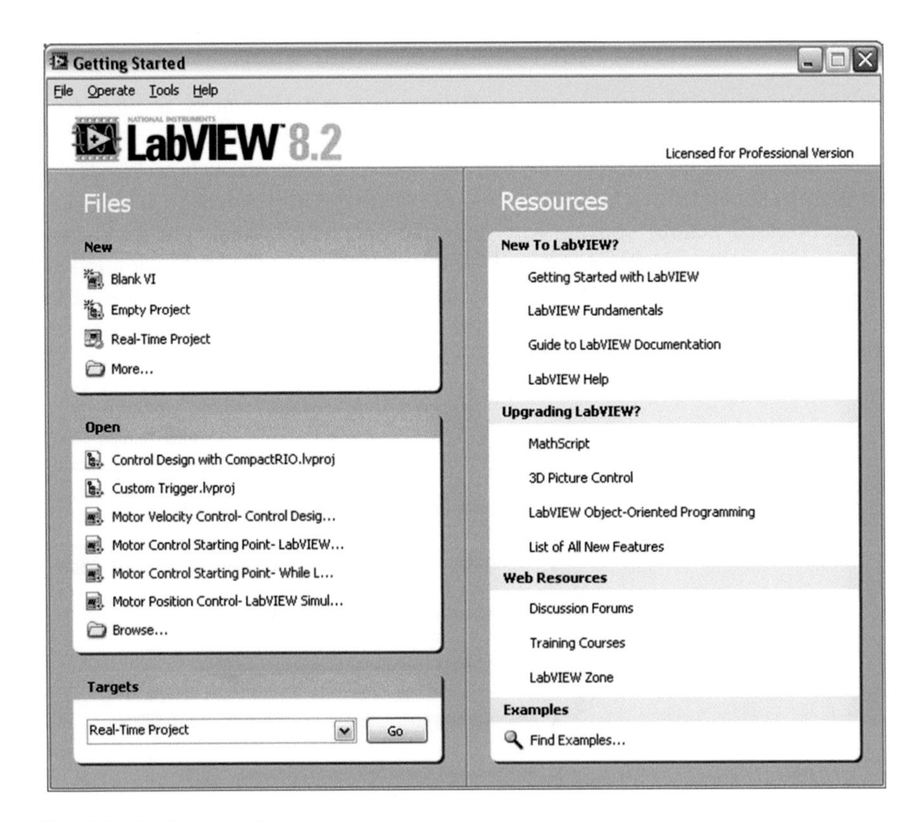

Fig. 2.13 Real-time project

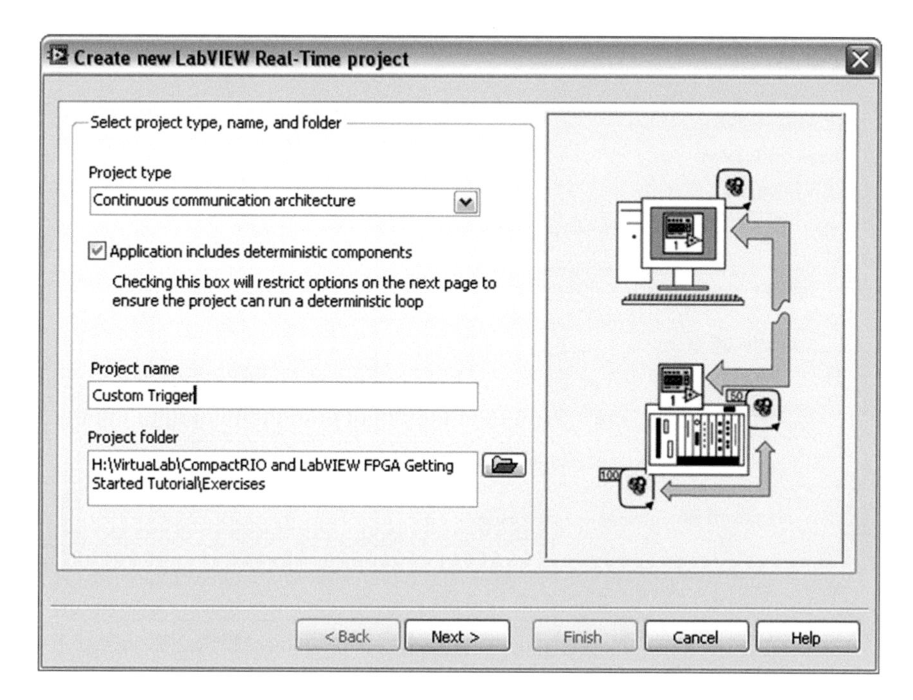

Fig. 2.14 Real-time folder

2. To select the working folder for your project, click the folder ( ) icon, navigate to **H:\VirtuaLab\CompactRIO and LabVIEW FPGA Getting Started Tutorial\Exercises**, and then click the **Current Folder** button. Name your project **Custom Trigger**. Keep all of the project defaults as shown below and click the **Next** button (see Fig. 2.14).
3. Change your **Target Configuration** to **Two loops**. Under the **Host Configuration** section (see Fig. 2.15), check the **Include user interface** box. Then click **Next**.
   The LabVIEW 8.20 Real-Time Project Wizard makes it easy to create a complete CompactRIO embedded system that includes an FPGA application (see Fig. 2.16), real-time processor application, and networked Windows host computer application. After this exercise is complete, you could use the template applications created by the wizard to create a complete networked system, including a deterministic loop running on the real-time controller to communicate with the FPGA and a lower priority loop to performed network communication, file logging, or additional analysis.
4. Click the **Browse** button to find the networked target you configured in MAX. Expand **Real-Time CompactRIO** folder and wait until your CompactRIO system is detected (see Fig. 2.17). Highlight your CompactRIO system and click **OK**. Then click **Next** to continue creating the real-time project.

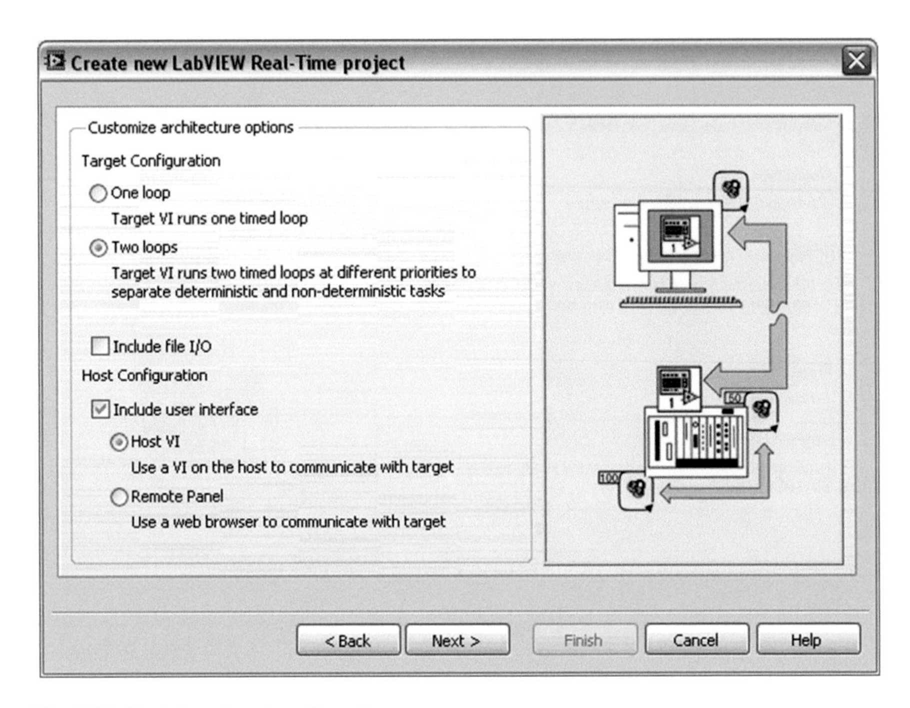

Fig. 2.15 Real-time target configuration

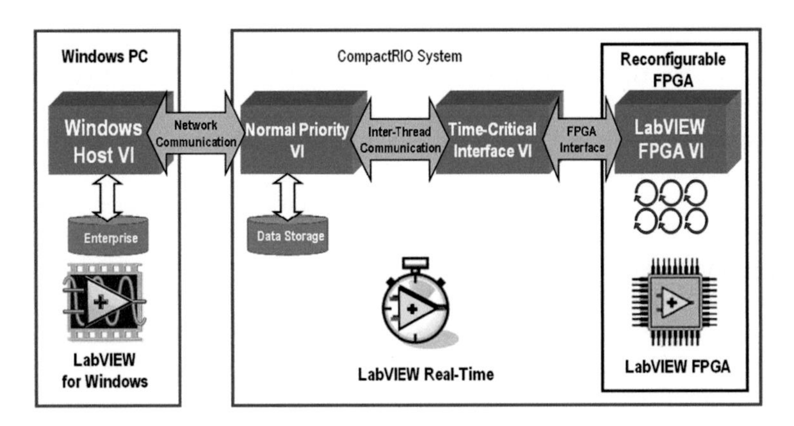

Fig. 2.16 Real-time application

5. Notice that the project wizard displays a preview of the project you configured. Click **Finish** to finalize the creation of the new real-time project and generate the application template code (see Fig. 2.18).
*When code generation is complete, two pre-built template applications will automatically open. The Windows host application (**host–network–RT (separate).vi**) includes a chart to plot the data sent by the CompactRIO system over*

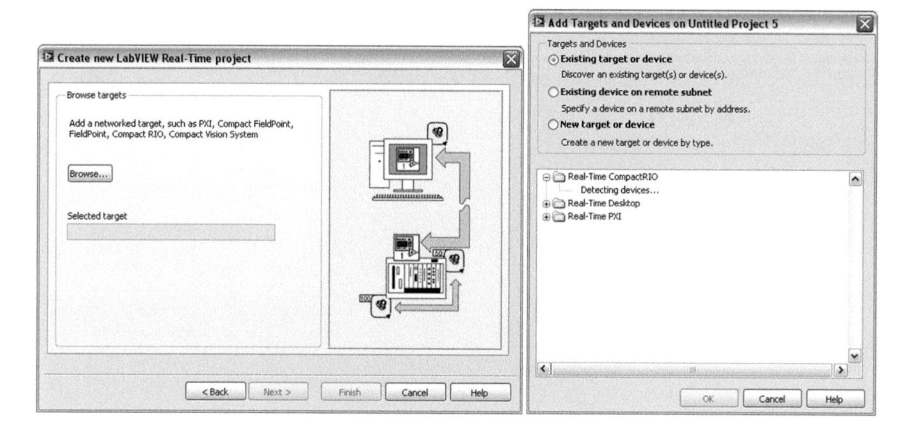

Fig. 2.17 CompactRIO detection

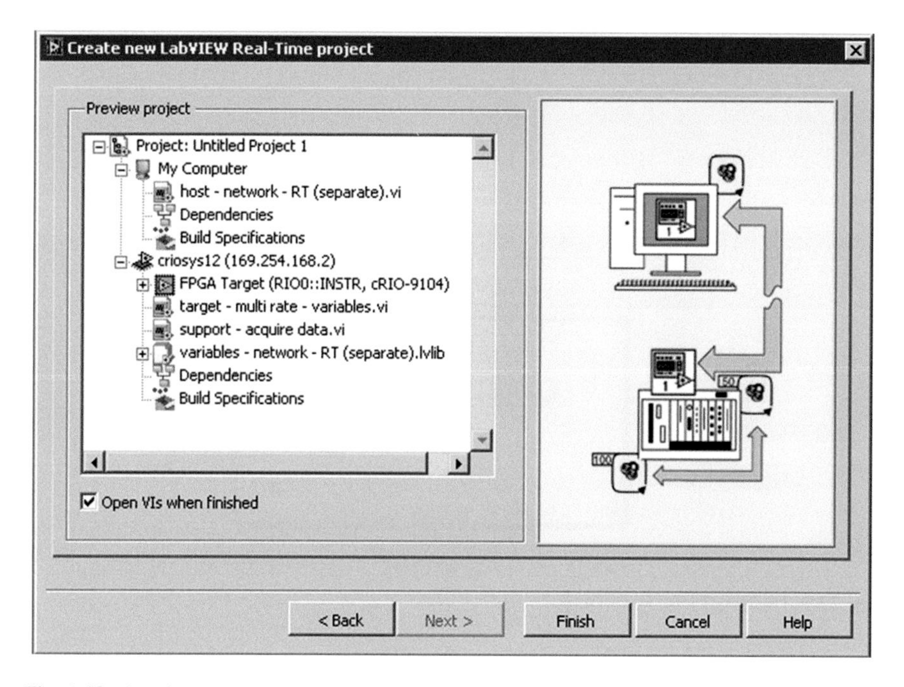

Fig. 2.18 Creating a new real-time project

the network, and a ***stop–network*** *shared variable that is used to halt execution of the real-time embedded application running on the CompactRIO system* (see Fig. 2.19).

6. In the **target–multi-rate–variables.vi** (see Fig. 2.20) real-time processor application, navigate to **Window≫Show Block Diagram**.

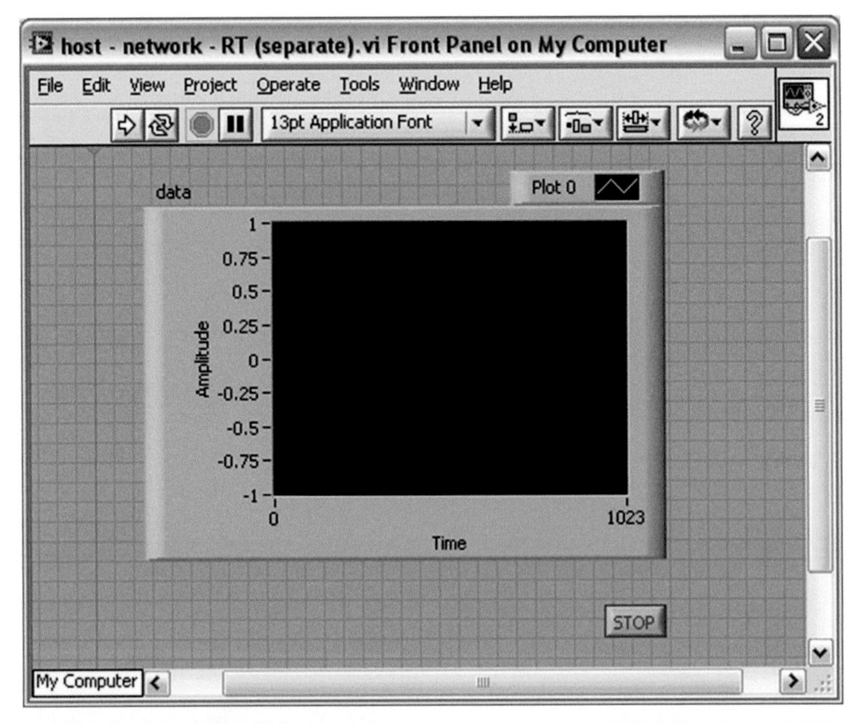

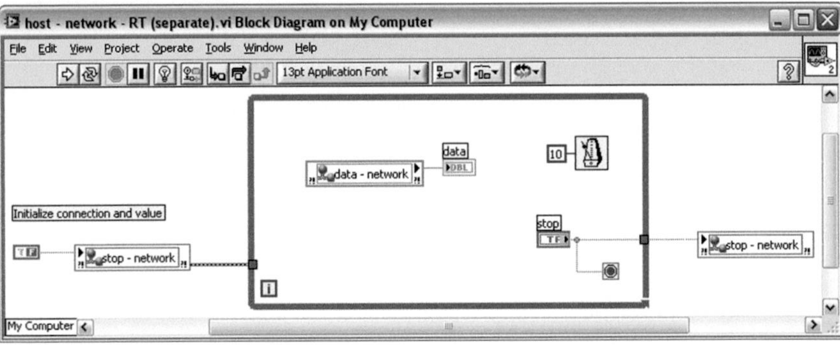

Fig. 2.19 Host–network–Real Time.vi

This embedded processor application produces a simulated I/O signal and sends the data to the Windows host computer using network-published shared variables. You would place any time critical routines, such as code to interface with your FPGA application within the top deterministic loop. Any lower priority non-deterministic tasks such as data logging or additional analysis would be placed in the bottom lower priority loop.

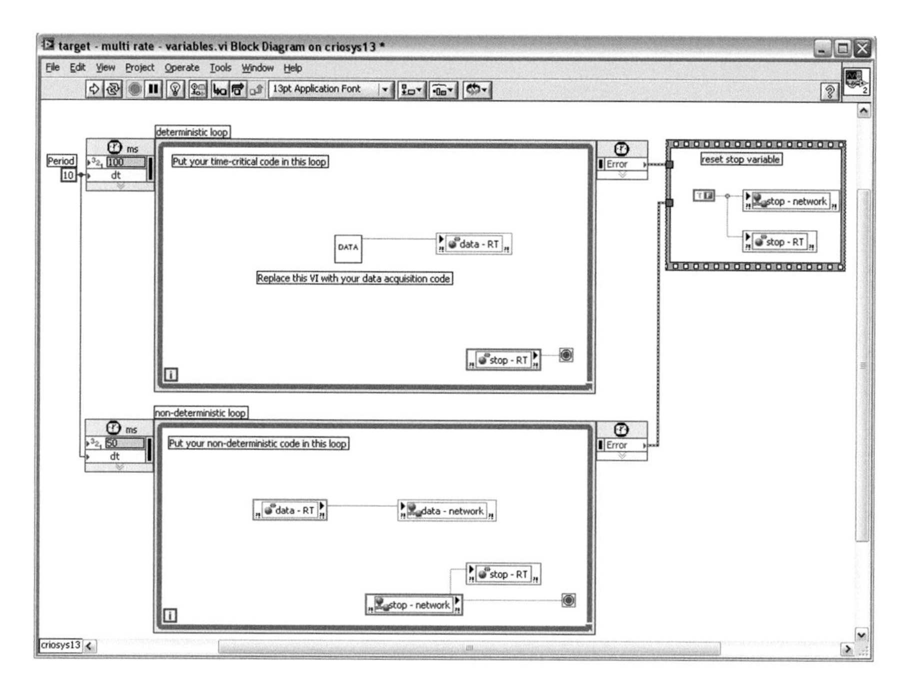

Fig. 2.20 Target–multi-rate–variables

7. Click the **Run** button on the real-time processor application (**target–multi-rate–variables.vi**). While the embedded application is being deployed, click the box next to **Close on successful completion** if it is not already checked (see Fig. 2.21).
8. Click the **Run** button on the Windows host application (***host–network–RT (separate).vi***). View the sinusoidal waveform displayed on host application chart (see Fig. 2.22). Click the **STOP** button on the host application and notice that the application stops running on both the host and real-time target.
9. In the **Project Explorer** window (see Fig. 2.23), right-click on the **FPGA Target** and select **New≫C Series Modules** to add your I/O modules to the project.
10. To automatically detect the I/O modules installed in your chassis (see Fig. 2.24), expand the **C Series Module** tab by clicking on the + symbol. Click **Continue** when the warning dialog window appears. A pre-built FPGA bit-stream will be downloaded to auto-detect the installed modules.

    *Note: If you are working offline without a network connection to your CompactRIO system, you can still develop your code by selecting **New target or device** and manually adding the I/O modules.*

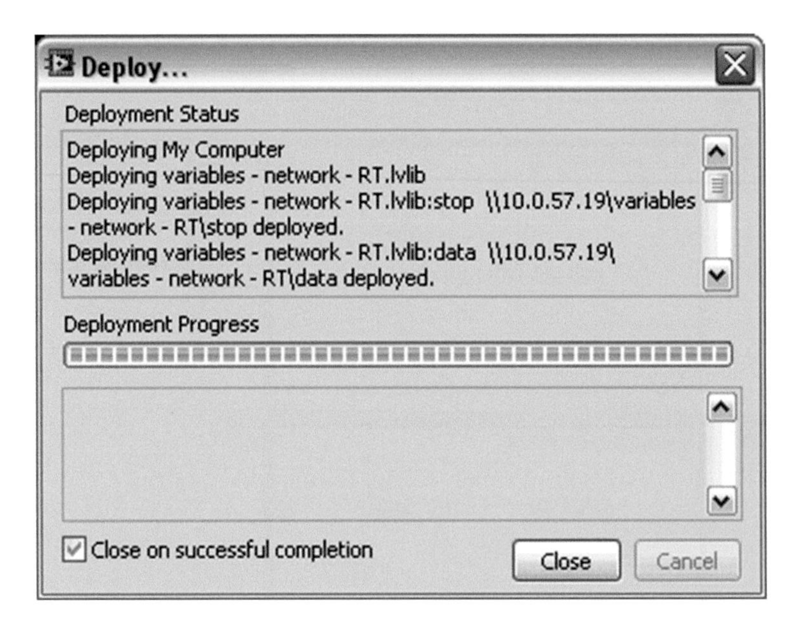

Fig. 2.21 Real-time program deployment

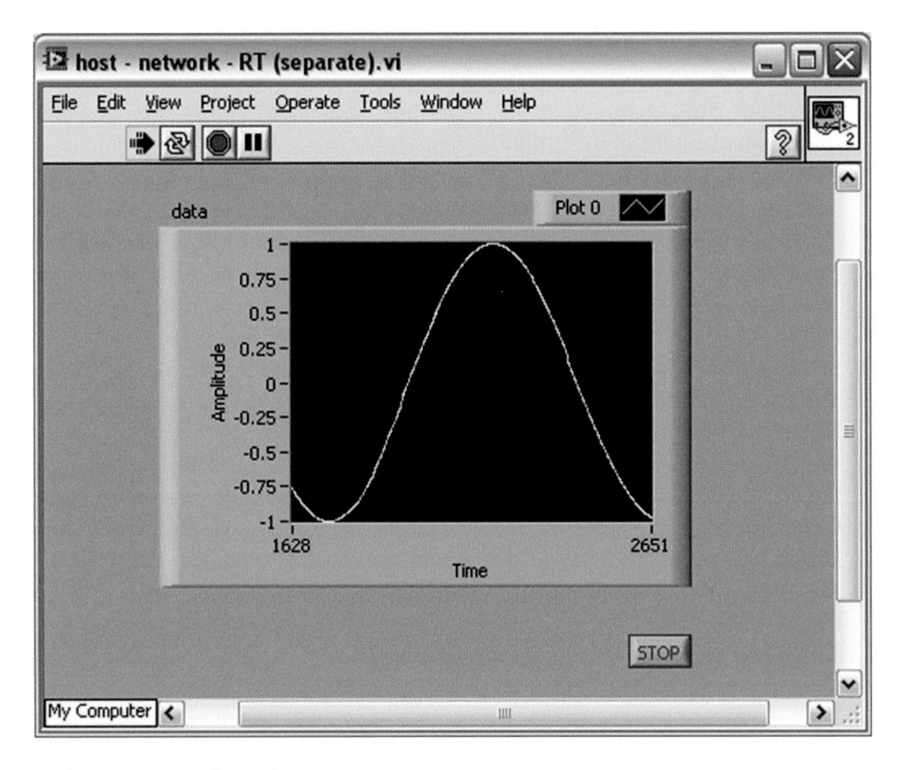

Fig. 2.22 Sine waveform displayed

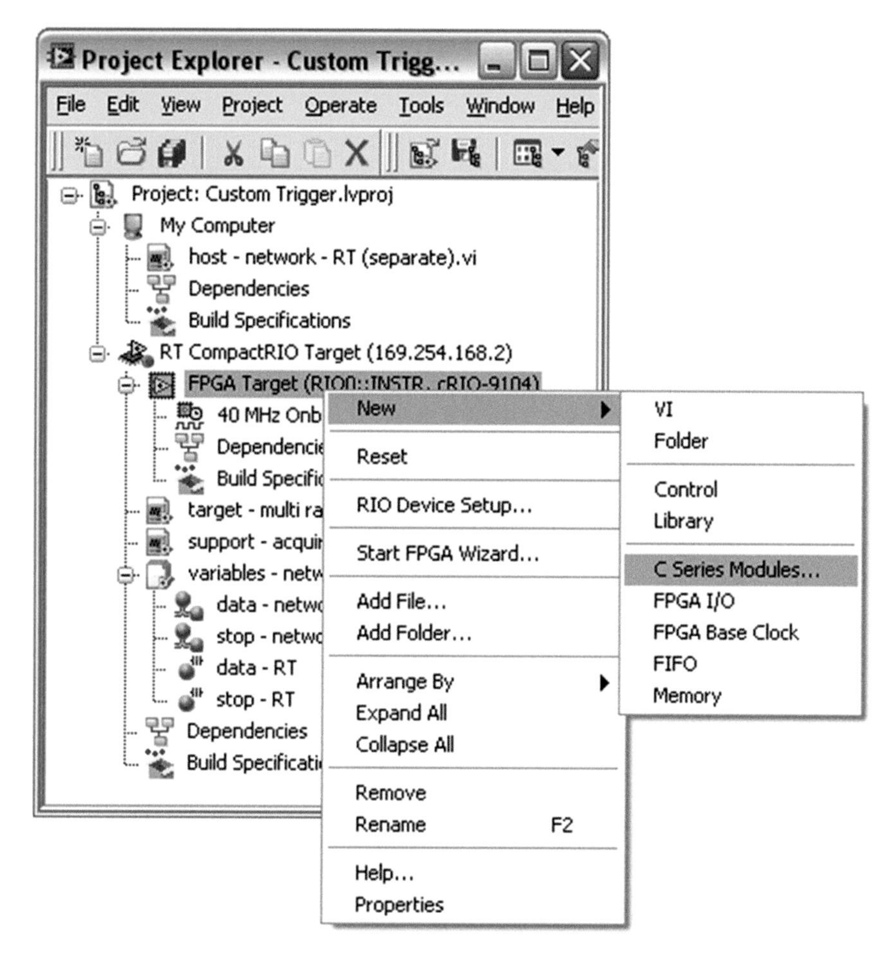

Fig. 2.23 Project explorer

11. After the modules are detected, select the modules that will be used in this exercise (see Fig. 2.25). To do this, first click on the **NI 9215** module, then hold down the **Ctrl** key and click on the **NI 9263** and **NI 9401** modules. Click **OK** to add all modules to your project.
12. In the **Project Explorer** window (see Fig. 2.26), right-click on the **FPGA Target** and select **New≫FPGA I/O** to add your I/O channels to the project. The **Analog Input** section is highlighted. To highlight all sections, hold down the **Shift** key and click on the **Digital Port Input and Output** section.
13. Next click the **Add** button to add all of the I/O channels. Then click the **OK** button to finish adding the channels to your project (see Fig. 2.27).

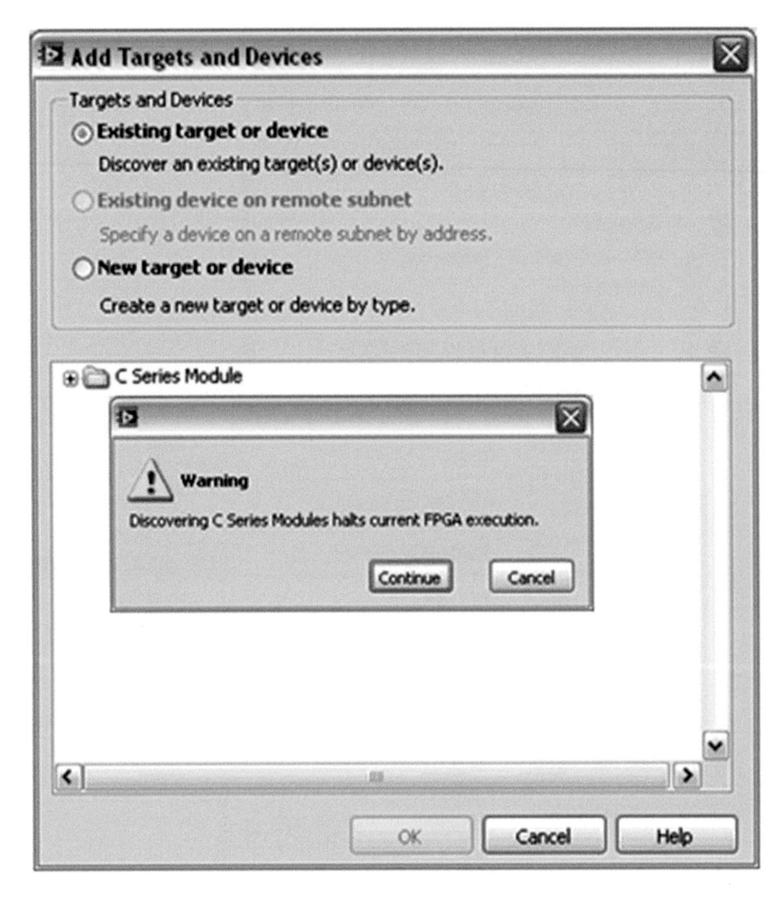

Fig. 2.24 Adding targets and devices

14. In the **Project Explorer** (see Fig. 2.28) window, right-click on the **FPGA Target** and select **Collapse All**. If you click the + symbol next to the **FPGA Target**, your LabVIEW Project should appear similar to what is shown below. Click the **Save All** button to save the project and all subVIs.

2.2 Developing the LabVIEW FPGA Application

One of the most important steps to build a fuzzy controller is to design an FPGA application, so the next section describes how an application is designed. When you finish a FPGA application, the front panel and block diagram of your completed FPGA application will look like Fig. 2.29.

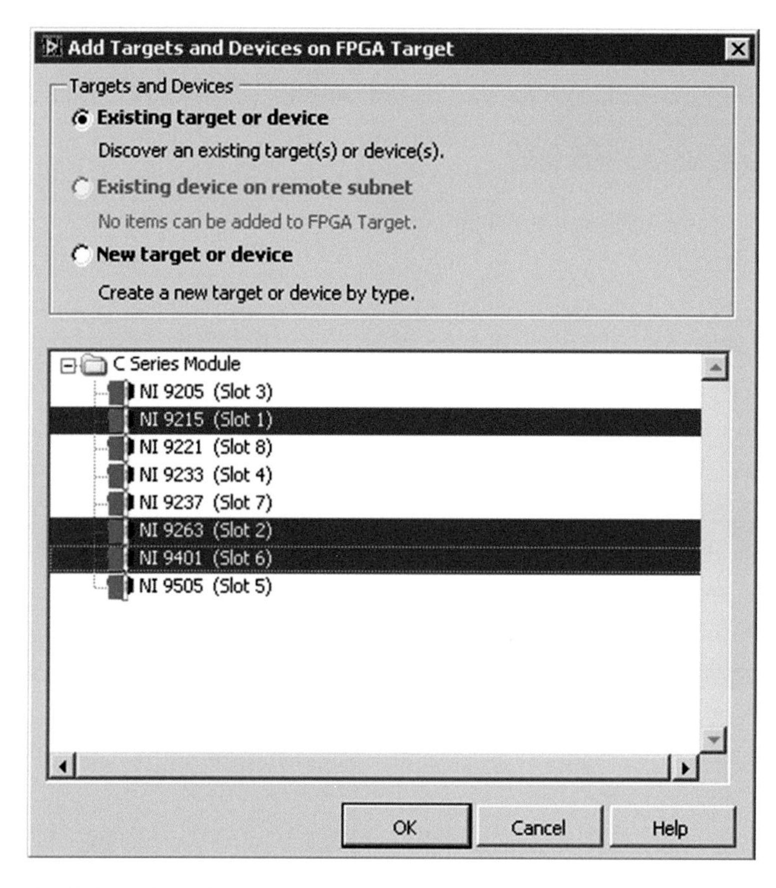

Fig. 2.25 Modules detected

The next step shows how an application is developed. *LabVIEW applications are called "Virtual Instruments" or "VIs."*

1. In the LabVIEW **Project Explorer**, right-click on the **FPGA Target** and select **New»VI** to start a new LabVIEW FPGA application. When the VI opens, navigate to **File»Save**. Then browse to the **H:\VirtuaLab\CompactRIO and LabVIEW FPGA Getting Started Tutorial\Exercises** folder and save the application as "**Simple AIAO (FPGA)**" (see Fig. 2.30).
   *To make it easier to distinguish the intended execution target, it is recommended that you include the words "**FPGA**" in the filename of your FPGA applications.*

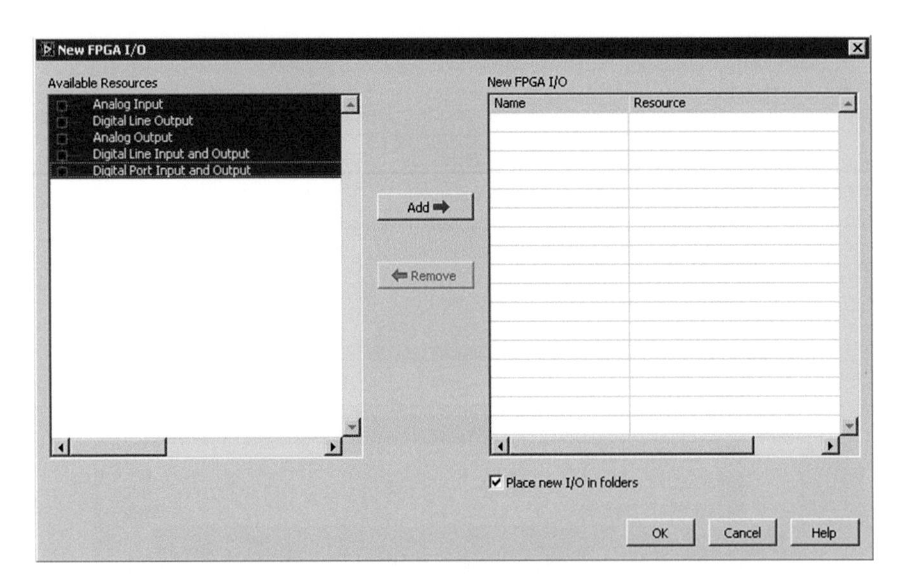

Fig. 2.26 FPGA I/O modules

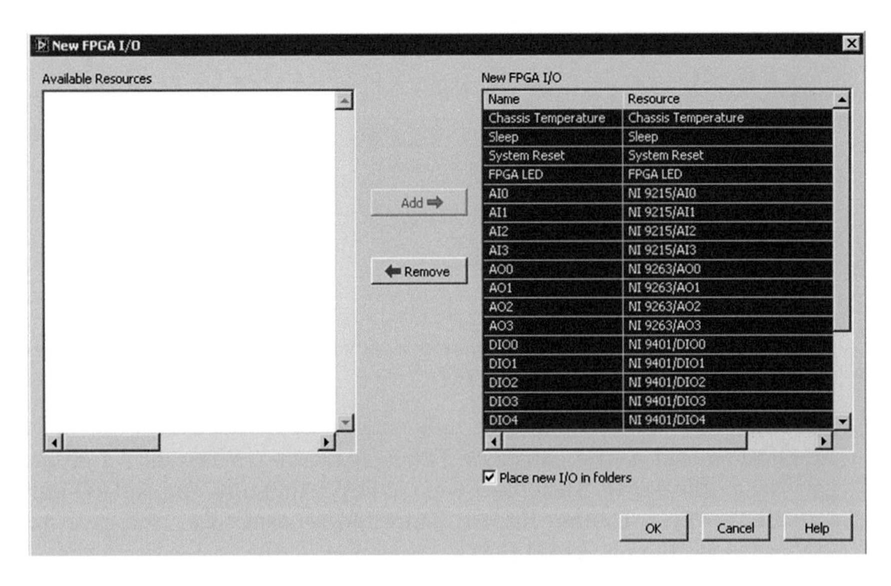

Fig. 2.27 Modules selected in the FPGA

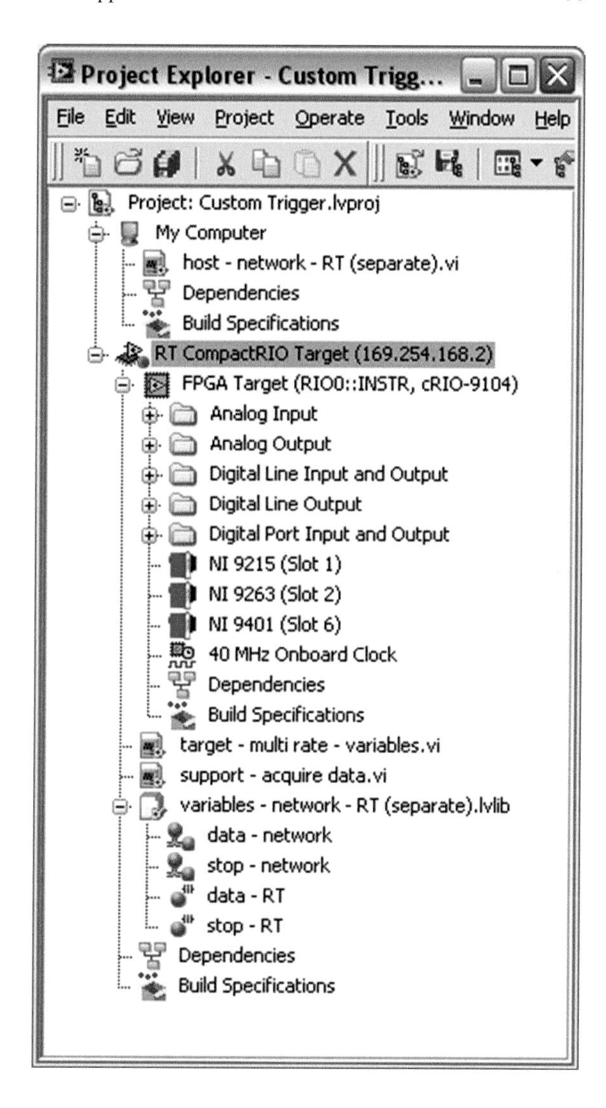

Fig. 2.28 Project explorer

2. Navigate to the block diagram window for your **Simple AIAO (FPGA)** application. (You can also make the block diagram appear by selecting **Window»Show block diagram** while viewing the front panel.) Right-click in the white area of the block diagram to display the **Functions** palette. Click on the thumb tack icon in the top left corner of the **Functions** palette to tack it down. Then navigate to the **Help** menu on your VI and select **Show context help**.

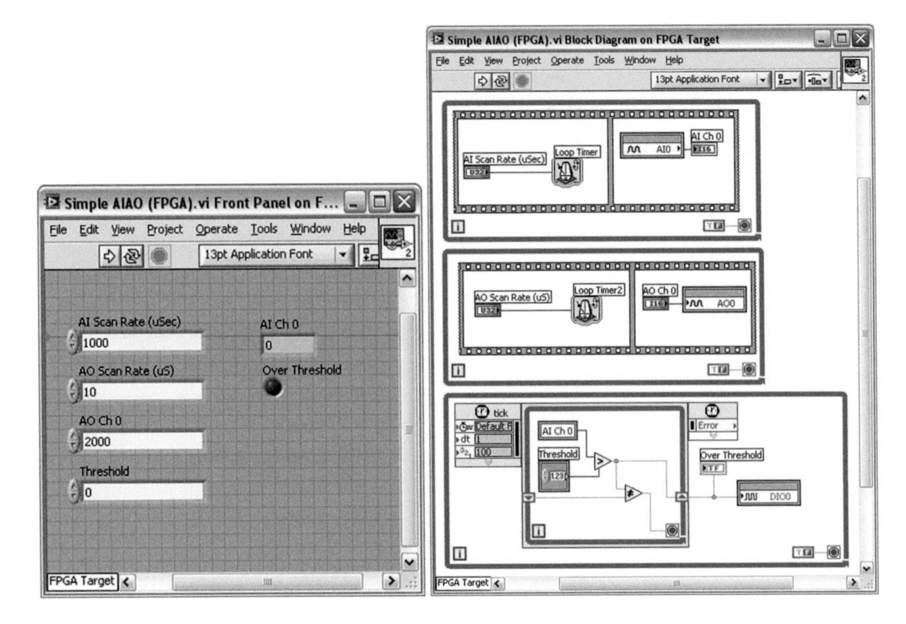

Fig. 2.29 Font panel and block diagram in FPGA application

3. Browse through the **Functions** palette to familiarize yourself with the many IP blocks that ship with LabVIEW FPGA. Be sure to browse through the **Programming Structures**, **Timing, Numeric, Boolean** logic**, Comparison, FPGA Math and Analysis, FPGA I/O**, and **Synchronization** palettes (see Fig. 2.31).
   If you have the NI SoftMotion Development Module installed, you will see an additional ***Spline Engine*** *function under the* ***VisionMotion*** *category that is used for high-performance multiaxis coordinated motion control. Other motion control IP blocks, such as a 32-bit motion PID controller function, are located in the LabVIEW examples directory. To locate these examples, navigate to* ***Help»Find Examples****. The NI Digital Filter Design Toolkit is another powerful add-on that provides the ability to generate your own custom signal processing IP blocks for LabVIEW FPGA. For fixed-point filter design, users can model quantization effects, optimize numeric representation/topology, and finally deploy the design using automatically generated LabVIEW FPGA code.*
   A common use for LabVIEW FPGA and RIO hardware is the development of custom triggering logic. In this exercise, you will program the FPGA to read in data from an analog channel, compare it to a threshold, and write a TRUE/FALSE value to a digital channel. If the analog input value exceeds the threshold then a NI 9401 digital output channel will turn on. You will use the indicator LED on the module and software front panel indicators on the FPGA application to view the status.

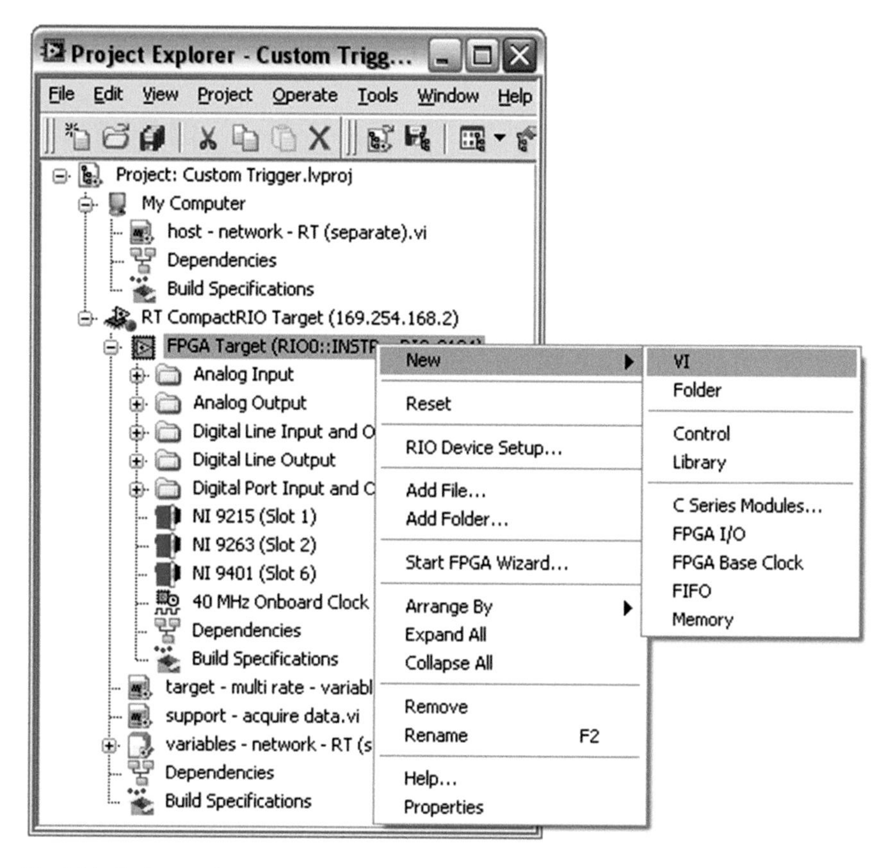

Fig. 2.30 Project explorer

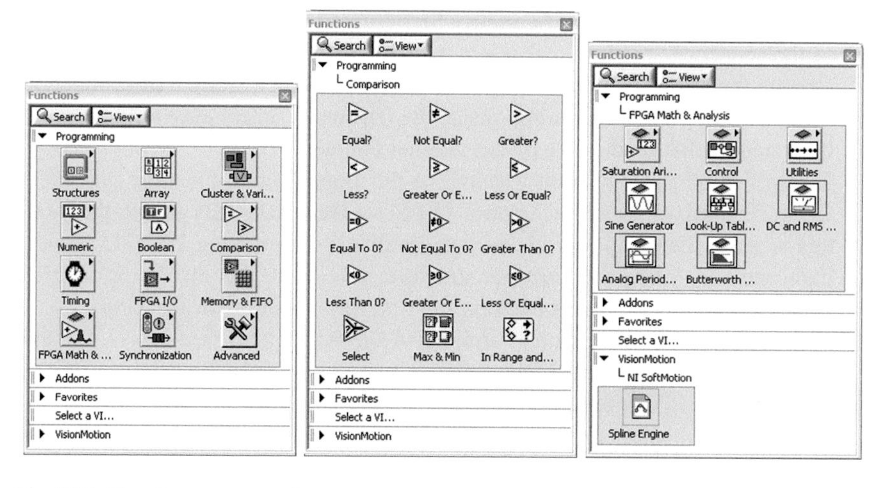

Fig. 2.31 Functions programming

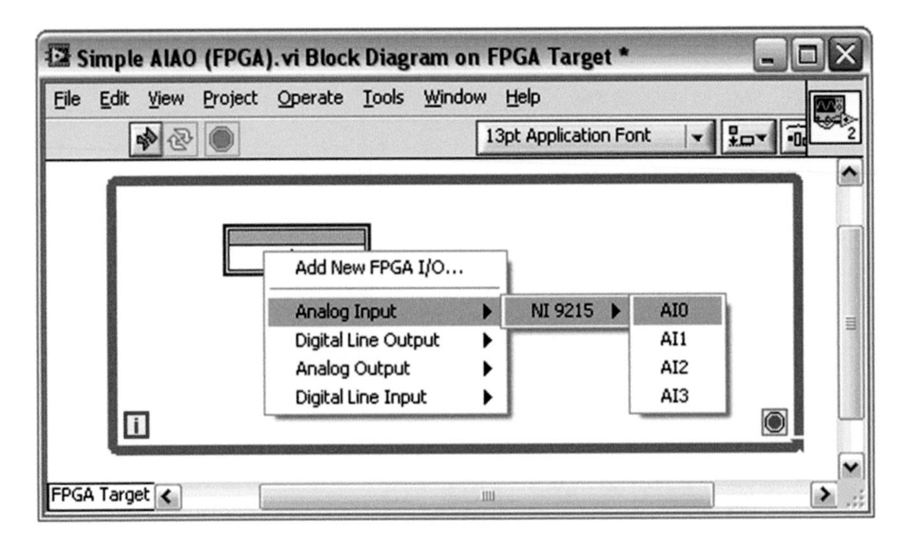

Fig. 2.32 Analog Input

4. We will begin building the application by reading in analog input channel 0 at a timed interval. First, place a **While Loop** from the **Functions»Structures** palette on the block diagram.
5. From the **Functions»FPGA I/O** palette, place a **FPGA I/O Node** function inside the **While Loop**.
6. Left-click on the **I/O Name** terminal and select **Analog Input»NI 9215»AI0** (see Fig. 2.32).
7. Right-click on the **AI0** terminal and select **Create»Indicator**. Label the indicator "**AI Ch 0**".
8. From the **Functions»Structures** palette enclose the **FPGA I/O Node** and the **AI Ch 0** indicator in a **Flat Sequence Structure**. Then right-click on the border of the sequence structure and select **Add Frame Before**.
9. Expand the left frame to make more room. Then place a **Loop Timer** function (**Functions»Time and Dialog**) inside the left frame. Select **µsec** as the counter units and **32 Bit** as the size of the internal counter.
10. Right-click on the left input terminal of the **Loop Timer** function and select **Create»Control**. Label the control "**AI Scan Rate (uS)**". By timing the loop, this will set the sampling rate of the simultaneous sampling NI 9215 analog input module. Using the sequence structure, you insure that the timing interval between samples is correct even on the first few iterations of the loop.
11. Right-click on the conditional terminal of the while loop (see Fig. 2.33) and select **Create»Constant**. Make sure the constant is set to the default value of **FALSE**. The FPGA application should appear as shown below.
    By placing the loop timing function in the first frame of the flat sequence structure, we ensure that the correct loop timing occurs on the first iteration of

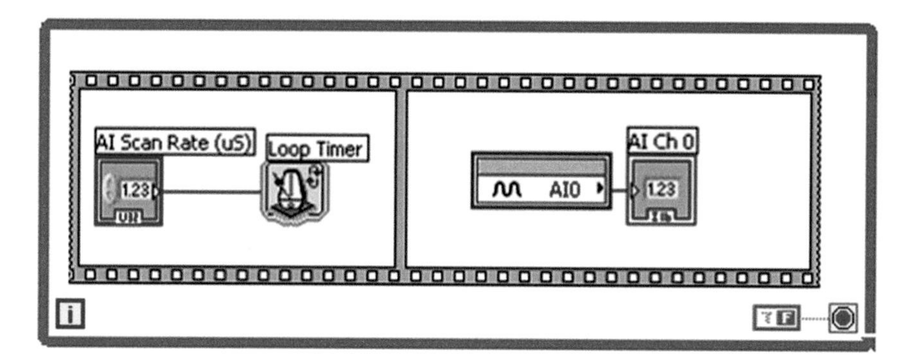

Fig. 2.33 While loop

the loop. If no sequence structure was used, the second acquisition would occur immediately after the first since no delay would be added. That's because on first execution, the **Loop Timer** *function sets its internal timing register but does not add a delay to the loop. In general, for any functions placed in parallel on the block diagram with no data dependencies, LabVIEW FPGA will synchronize the start of each parallel function.*

12. Place another **While Loop** structure (**Functions»Structures**) on the block diagram below the analog input loop you just created.
13. Drop down an **FPGA I/O Node** function (**Functions»FPGA I/O**). Left-click the I/O node and select **Analog Output»NI 9263»AO0** to access the analog output channel 0.
14. Right-click on the **AO0** terminal and select **Create»Control**. Label the control "**AO Ch 0**".
15. Following the same process as you did before, enclose the **FPGA I/O Node** in a **Flat Sequence Structure**, add a frame to the left, and drop in a **Loop Timer** function. Select **μsec** as the counter units and **32 Bit** as the size of the internal counter.
16. Right-click on the input terminal to the **Loop Timer2** function and select **Create»Control**. Label the control "**AO Scan Rate (uS)**." This control will set the update rate of the NI 9263 analog output module.
17. Right-click on the conditional terminal of the while loop and select **Create»Constant**. Make sure the constant is set to the default value of **FALSE**. The FPGA application should appear as shown below in Fig. 2.34. *Note that in LabVIEW FPGA, each loop will execute in parallel. Unlike processors, FPGAs use dedicated hardware for processing logic and do not have an operating system. FPGAs are truly parallel in nature so different processing operations do not have to compete for the same resources. As a result, the performance of one part of the application is not affected when additional processing is added. Also, multiple loops can run on a single FPGA device at different rates. To learn more, view the FPGA-based Control FAQ.*

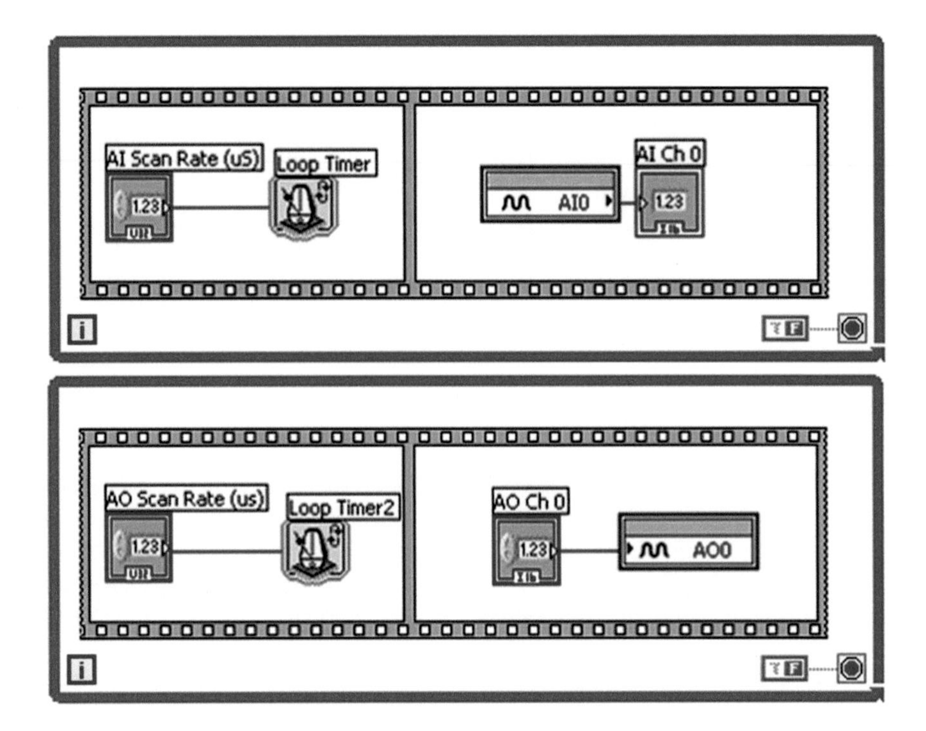

Fig. 2.34 Condition terminal while loop

18. Next we will add another loop to perform custom triggering and digital output. Drop down an additional **While Loop** below the first two loops.
    In this application, we will look for a trigger condition to see if the analog input is above or below the threshold. This custom trigger logic is implemented inside a SCTL, which will execute at a 40 MHz rate. If a change is detected, we will set a digital output on the NI 9401 module to the appropriate value. The NI 9474 module has a worst-case output propagation delay of 1 μs.
19. Place a **Timed Loop** (**Functions**≫**Structures**≫**Timed Structures**) inside the while loop.
20. Right-click on the **AI Ch 0** indicator and select **Create**≫**Local Variable**. Place the local variable inside the **Timed Loop** structure. Right-click on the local variable and select **Change To Read**.
21. Drop a **Greater?** function (**Functions**≫**Comparison**) on the block diagram and wire the **AI Ch 0** local variable to the top input terminal. Then right-click on the lower terminal and select **Create**≫**Control**. Label the control "**Threshold**."
22. Right-click on the border of the **Timed Loop** and select **Add Shift Register**. This register will store the value of the comparison function and pass it from one iteration of the loop to the next. Connect the output of the **Greater?** function to the input of the shift register on the right border of the timed loop.

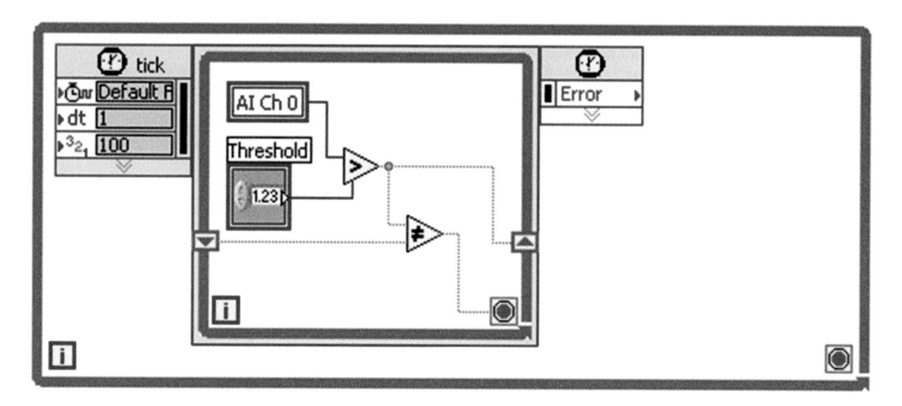

Fig. 2.35 Time loop

23. Place a **Not Equal?** function (**Functions**»**Comparison**) to the right of the **Greater?** function and connect the green signal wire to the top terminal of the **Not Equal?** function. Then wire the Shift Register value from the left side of the timed loop to the bottom terminal the **Not Equal?** function. Finally, wire the output of the **Not Equal?** function to the conditional terminal of the timed loop. If the analog in value has risen above or dropped below the threshold value then we will stop the timed loop and update the digital output (see Fig. 2.35).
    The SCTL structure instructs the LabVIEW compiler to execute the code inside of it within a single 25 ns clock cycle of the FPGA (40 MHz). Code inside of a SCTL not only executes faster, but also uses fewer FPGA resources or "slices." However, certain functions are not supported in the SCTL such as analog input and analog output I/O nodes and the ***Quotient and Remainder*** *function (an integer math divide function). If you use an unsupported function inside of a SCTL, you will get a compile error early in the compilation process.*
24. Right-click the output of the **Shift Register** on the right border of the timed loop and select **Create**»**Indicator**. Name this indicator "**Over Threshold**."
25. Drop down an **FPGA I/O Node** (**Functions**»**FPGA Device I/O**) next to your **Over Threshold** indicator (outside of the timed loop). Left-click the **FPGA I/O Node** and select **Digital Line Input and Output**»**NI 9401**»**DIO0** to access the digital output channel 0. Right-click on the **FPGA I/O Node** and select **Change to Write**. Then wire the output signal from the right shift register to the **Digital Output** node.
26. Wire a **FALSE Boolean Constant** to the Loop Condition terminal of the outer while loop. The FPGA application should appear similar to that shown below in Fig. 2.36.

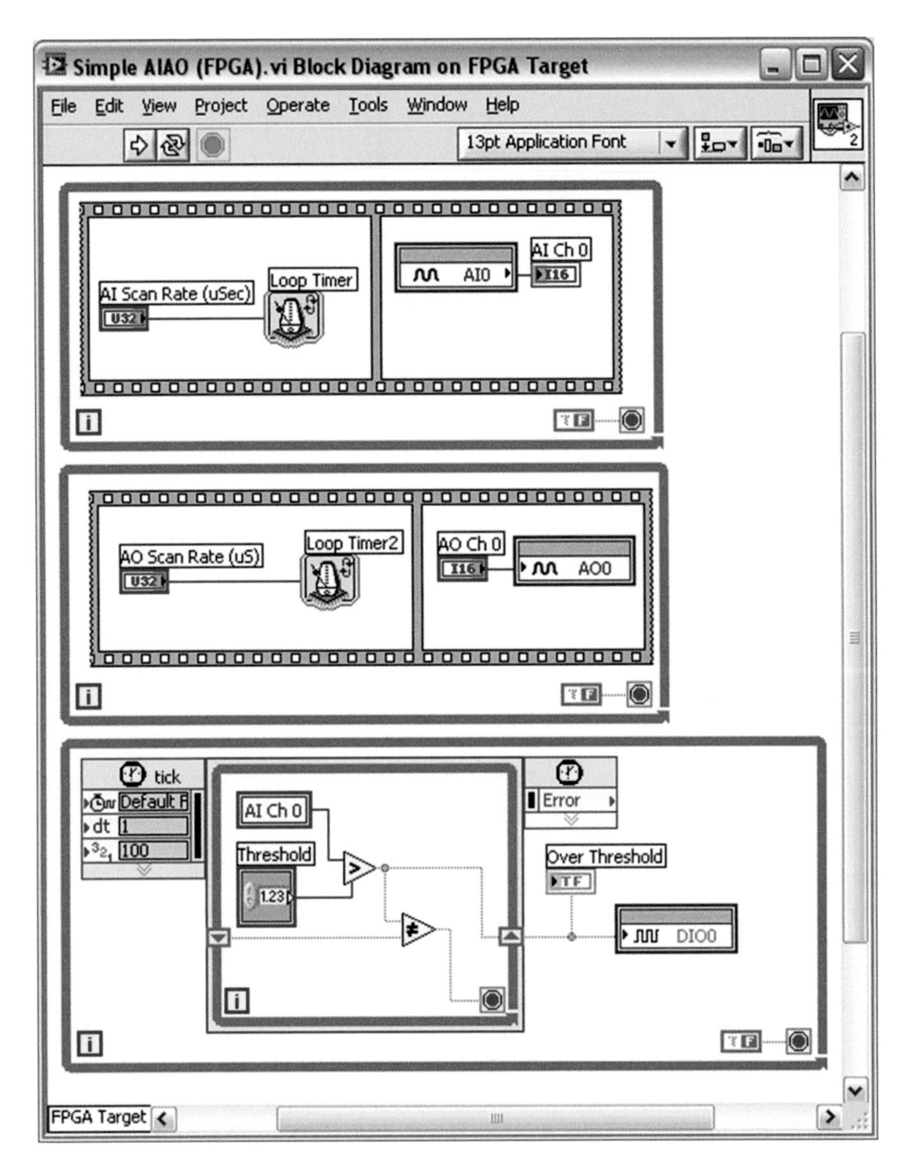

Fig. 2.36 Loop conditional terminal

27. Navigate to the front panel of your FPGA application and arrange the controls and indicators as shown in Fig. 2.37. Set the **AI Scan Rate (uS)** to **1000**. Set the **AO Scan Rate (us)** to **10**. Set **AO Ch 0** to **2000**. Then navigate to the **Edit** menu and select **Make current values default**.
    This will set the FPGA application so that at startup the analog output loop runs at 100 kS/s (10 μs) and the analog input loop runs at 1 kS/s (1000 μs).

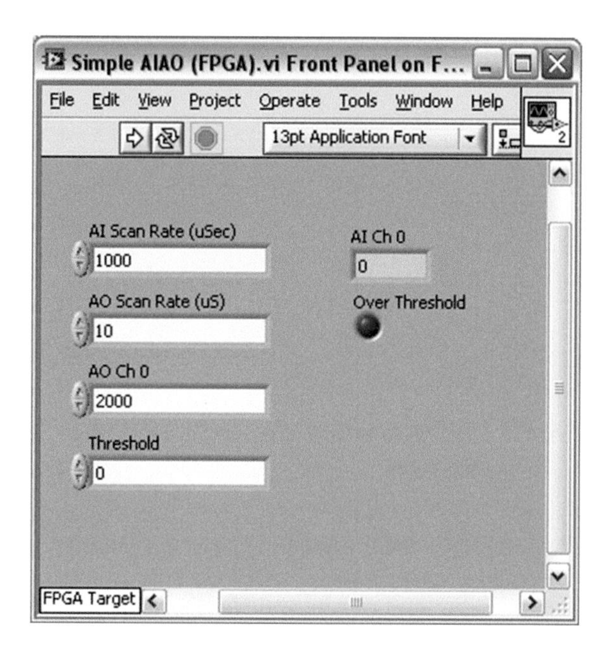

Fig. 2.37 FPGA applications front panel

*Setting the **Threshold** to 0 and the NI 9263 analog output value (**AO Ch 0**) to 2000 (about 0.653 V) will cause the digital output to go high due to an over threshold condition when the application starts. For more information about the scaling from integer values to voltage outputs on the NI 9263, refer to the NI 9263 Operating Instructions manual.*

28. When you are ready to compile navigate to **File**≫**Save All** to save all open applications and the project.

2.3 Compiling the FPGA Application

In this section you will compile the LabVIEW FPGA application and learn more about the compilation process and view the compilation report.

1. Click the **Run** button to start the compile process.
2. Sit back and enjoy 5–10 min of relaxation while your LabVIEW FPGA application compiles. To better understand the LabVIEW FPGA compilation process, review the information below.

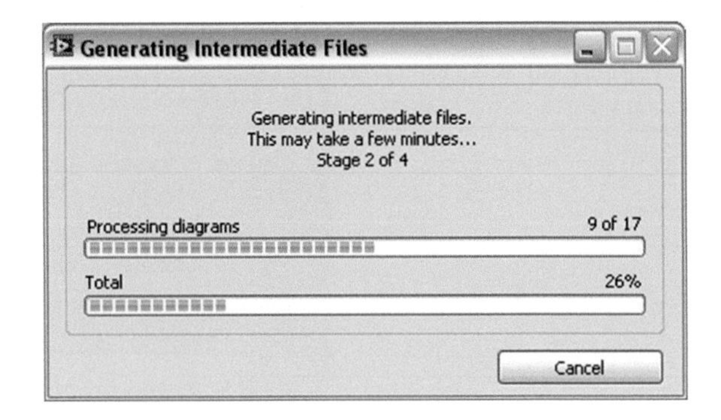

Fig. 2.38 FPGA compilation process

2.3.1 Understanding the LabVIEW FPGA Compilation Process

The LabVIEW FPGA Module [7] *uses an industry standard Xilinx ISE compiler. First, your graphical LabVIEW FPGA code is translated to text-based VHDL code. At this time, the* ***Generating Intermediate Files*** *dialogue is displayed* (see Fig. 2.38).

Then the Xilinx ISE compiler tools are invoked and the VHDL code is optimized, reduced, and synthesized into a hardware circuit realization of your LabVIEW design. This process also applies timing constraints on the circuit design that ensure an efficient use of FPGA resources (sometimes called "fabric").

A great deal of optimization is performed during the compilation process to reduce digital logic and create an optimal implementation of the LabVIEW application. The end result is a highly optimized silicon implementation that provides true parallel processing with the performance and reliability benefits of dedicated hardware circuitry. Since there is no operating system on the FPGA chip, the code is implemented in a way that ensures maximum performance and reliability (see Fig. 2.39).

The end result is a bit stream file that is loaded into your LabVIEW FPGA.VI file. When you run the application, the bitstream is loaded into the FPGA chip to configure the gate array logic. While the application is running on the FPGA, data for the front panel controls and indicators is passed over the network several times per second to enable ***Interactive Mode*** *testing of the application. The update rate in interactive mode communication is typically limited to about 10 S/s. Later you could build a real-time host interface to the FPGA application that enables high-speed data transfer and interrupt synchronization between the floating-point host processor and integer-based FPGA chipset.*

The diagram shown in Fig. 2.40 *is a summary of the LabVIEW FPGA compilation process. To learn more, click to view an application note on FPGA-based control.*

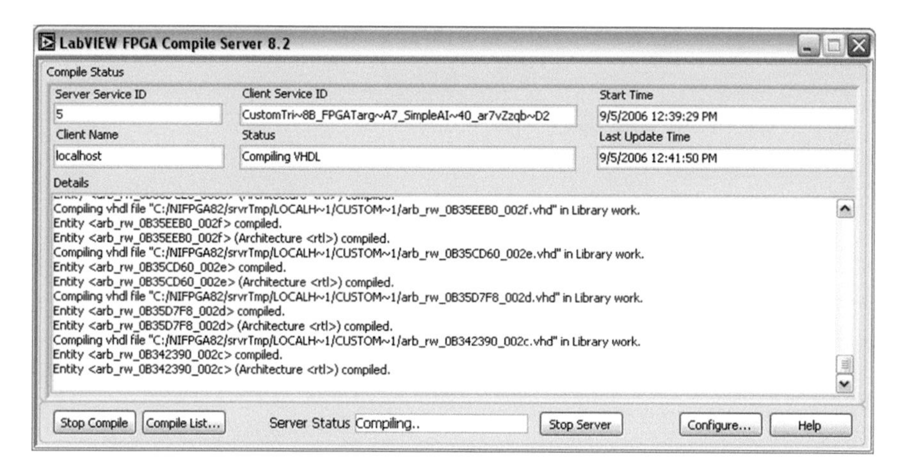

Fig. 2.39 FPGA compilation server

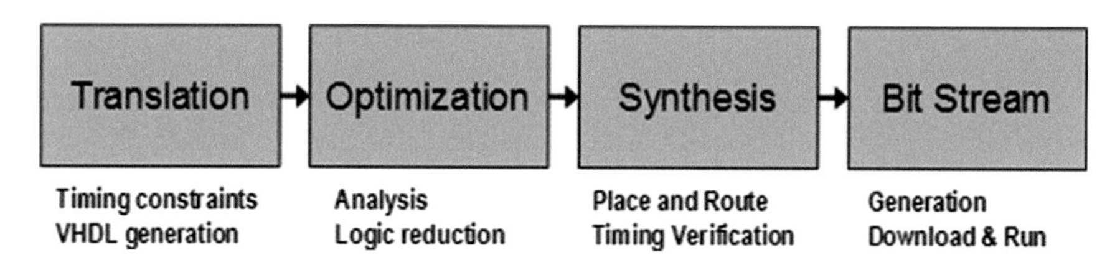

Fig. 2.40 Summary of the LabVIEW FPGA compilation process

2.3.2 FPGA Clock Speed

By default, the FPGA clock runs at 40 MHz. This means that one ***Tick*** *of the FPGA clock is equal to 25 ns. By changing the compile options, you can increase the FPGA clock speed up to 200 MHz (5 ns). There are some drawbacks to using higher clock speeds that you should be aware of before changing the compile option. For more information, refer to the CompactRIO Technical Developers Library by visiting* [8] *or by right-clicking on the* ***40 MHz Onboard Clock*** *item in the project and selecting* ***Help****.*

2.3.3 The Compilation Report

When the compilation is complete, the compile report will be generated. This report shows the start and end compilation time, the number of SLICEs used, a compiled clock rate (40 MHz), and an estimated maximum clock rate (see Fig. 2.41).

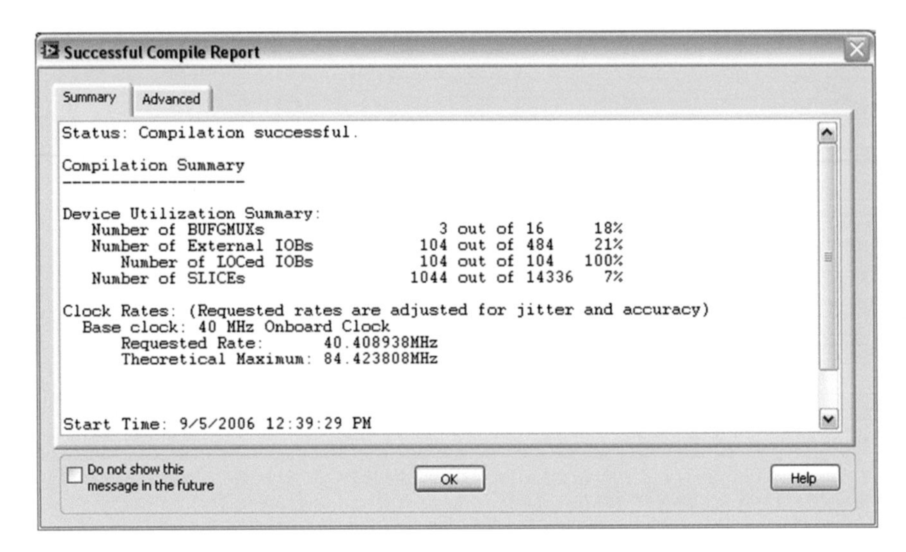

Fig. 2.41 Compilation report

A SLICE is a collection of logic components on the FPGA. The percentage shown is the percentage of the FPGA used. In most cases, you can actually fit more onto the FPGA and run the code faster than this report would lead you to believe. For small applications, the compiler does not "try as hard" to optimize, as long as the timing and other design constraints are met. As the FPGA reaches greater than 90 % usage, the compiler performs heavy optimization to make the most efficient use of resources.

2.4 Advanced Methods for LABVIEW FPGA

This section covers a number of advanced tips and tricks to cut your development time when creating high-performance control systems with LabVIEW FPGA and CompactRIO. It will be introduced the debugging techniques such as simulation that will make you confident before you hit the compile button. You will also learn a number of recommended programming practices, how to avoid common mistakes, and numerous methods to create fast, efficient, and reliable LabVIEW FPGA applications. Throughout this section, we will be walking you a number of examples that were developed to create a high performance control system for a brushed DC motor. We will be showing you a variety of the programming techniques that were used in the creation of LabVIEW FPGA subVIs for generating the PWM Drive signal, decoding the digital pulses from the Quadrature Encoder sensor, and performing PID control to close the motor position loop. The end result is a high performance control system with sub-nanosecond timing jitter, multiple 40 MHz processing loops, and that consumes only 17 % of a 3 million gate FPGA.

2.4.1 Introduction

To help you understand the role of the FPGA in a typical CompactRIO control system application, first we will review a typical software architecture (see Fig. 2.42).

First, the FPGA provides an interface to the I/O modules using an elemental I/O node interface. In some cases, the I/O is as simple as reading the voltage from an analog input module. For more complicated I/O types like the Quadrature encoder sensors that are common in motor control, the FPGA performs additional processing to convert raw digital signals into a meaningful measurement, such as the position and speed of a motor. In addition to I/O, the FPGA is commonly used for analog PID control, digital true/false logic, and event response.

Sending data from the FPGA to the real-time processor is as simple as creating a control or indicator on the front panel of the LabVIEW FPGA application. For high-speed buffered data, you can use DMA to stream data from FPGA memory to processor memory. The FPGA can also generate interrupts, which cause lower priority tasks to be interrupted on the host processor. This provides a way for the FPGA to synchronize the execution of code on the host processor, which can then perform calculations and respond in a deterministic fashion. In general, the real-time processor is slower than the FPGA but offers an extensive palette of floating-point control, math, and signal processing functions.

The CompactRIO processor executes a multithreaded, hard real-time operating system and is programmed using LabVIEW Real-Time. By multithreaded, we mean that it can execute multiple pieces of code, or loops, at different priorities. That

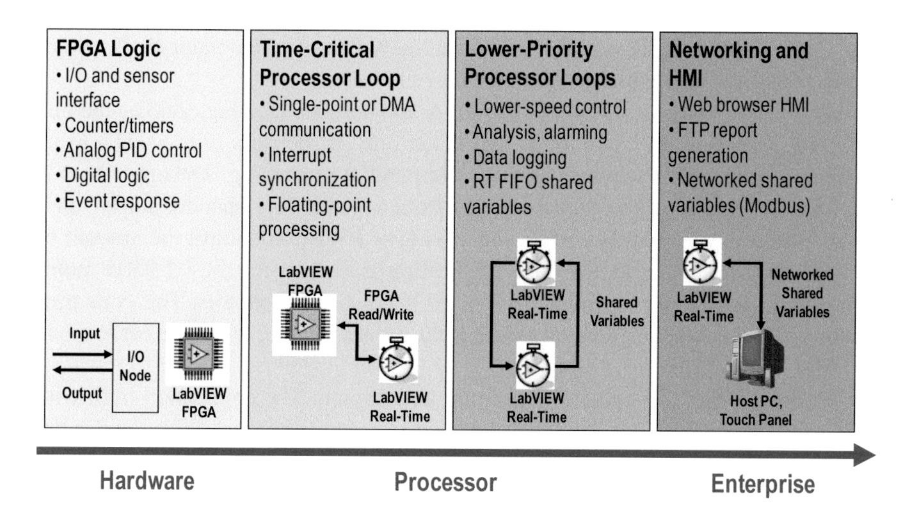

Fig. 2.42 Typical software architecture

means that you can add functionality to your application, such as data logging, lower speed control loops, or alarming without interfering with higher priority tasks that are more time sensitive. You can share data between loops of different priorities without causing interference using RT FIFO shared variables.

While the CompactRIO system is capable of standalone operation, many applications involve networking and a human–machine interface, or HMI. The two easiest ways to communicate with your CompactRIO system are through the web browser HMI or through file transfer protocol, or FTP. CompactRIO features a built-in web server that can host the front panel user interface of the lower priority loops in the embedded application. Alternately, you can use the LabVIEW Touch Panel module to create a low-cost HMI for your system. CompactRIO also has a built-in Modbus server that can publish or receive data from networked devices such as PLCs. Modbus/TCP is one of the most commonly used industrial networking protocols over Ethernet.

Let us take a look at five key development techniques that will help you create reliable and high-performance LabVIEW FPGA applications.

2.4.2 Technique 1: Use Single-Cycle Timed Loops (SCTLs)

The first development technique we will introduce is the use of SCTLs, in LabVIEW FPGA.

SCTLs work by telling the LabVIEW FPGA compiler to optimize the code inside, and add the special timing constraint that the code must execute in a single tick of the FPGA clock. Code compiled inside a SCTL is more optimized and takes up less space on the FPGA compared to the same code inside of a regular while loop. Code inside a SCTL also executes extremely fast. At the default clock rate of 40 MHz, one cycle is equal to just 25 ns.

Below are two identical LabVIEW FPGA applications (see Fig. 2.43)—the one on the left uses normal while loops, while the one on the right uses SCTLs in its subVIs. This example shows off the power of parallel processing. The upper loop is reading and processing the digital signals from a quadrature encoder sensor on a motor and the lower loop is performing PWM, or PWM, to control the amount of power being sent to the motor. This application is written for the NI 9505 motor drive module which can control up to 8 A, 30 V brushed DC motors. This code runs extremely fast—in the application on the right we are running two different loops at a 40 MHz clock rate.

The results from our compile report are also shown. The application built with SCTLs uses fewer SLICEs, but it takes longer to compile because the compiler has to work harder to meeting the timing constraints applied by the SCTL.

Now let us take a look at how the SCTL works in more depth.

When code is compiled in a normal while loop, LabVIEW FPGA inserts flip-flops to clock data from one function to the next, thereby enforcing the synchronous dataflow nature of LabVIEW and preventing race conditions. The

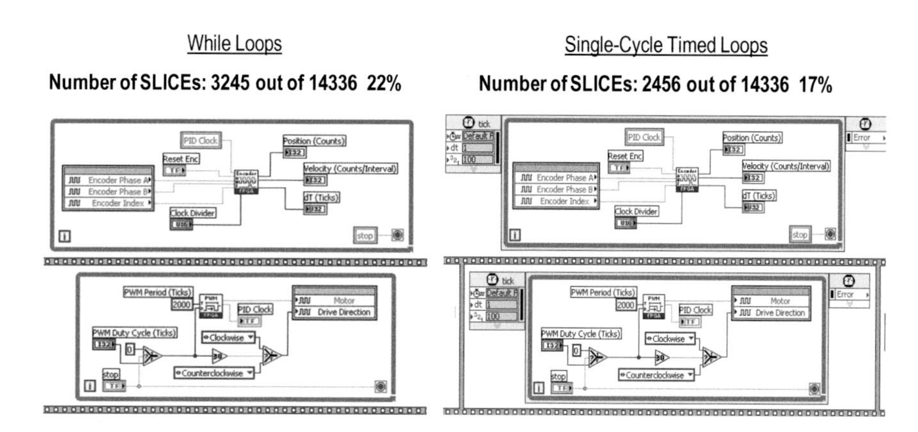

Fig. 2.43 Two loop cycles

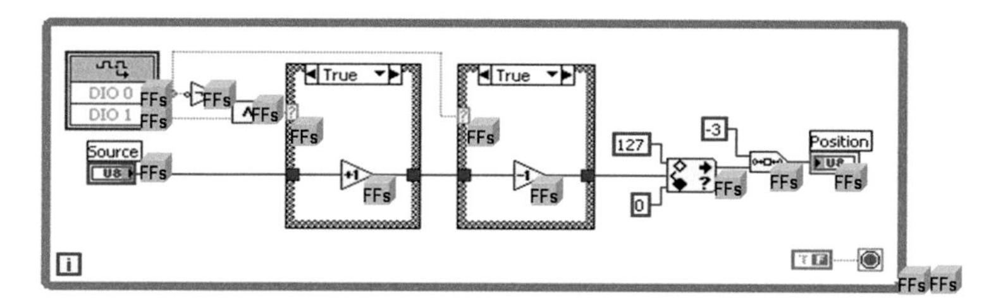

Fig. 2.44 Single-cycle time loop

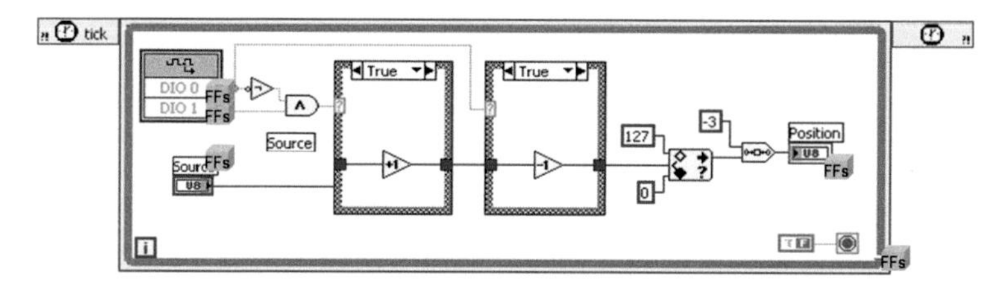

Fig. 2.45 Single-cycle time loop only I/O have flip-flops

flip-flops are marked here with the FF boxes drawn at the output of each function (see Fig. 2.44).

Below is the same code compiled into a SCTL. Here you see that only the inputs and outputs of the loop have flip-flops (see Fig. 2.45). The internal code is implemented in a more parallel fashion and more logic reduction is done to optimize the code in between the inputs and outputs of the loop.

As you can see, SCTLs are a simple way to optimize your LabVIEW FPGA code. So what is the catch? Why would not you always use the SCTL? There are some limitations to the use of SCTLs as it is shown in Table 2.2.

Table 2.2 Single cycle time loop limitations

Items not allowed in SCTL	Suggested alternative
Long sequences of serial code	Make the code more parallel. Insert feedback nodes in the wires to add pipelining
Quotient and remainder	Use a scale by power of 2 to do integer division, or use the fixed-point math library
Loop timer, wait functions	Use a tick count function to trigger an event instead
Analog input, analog output	Place in a separate while loop and use local variables to send data
While loops	For nested subVIs, use feedback nodes to hold state

To use the SCTL all operations inside the SCTL must fit within one cycle of the FPGA clock. In the beginning of the compile process, the code generator will give an error message if the SCTL cannot generate the proper code for the compiler. That means that long sequences of serial code may not be able to fit in a SCTL. By serial code, we mean code where the results of one calculation are needed by the next operation, preventing the calculations from being executed in parallel. To fix this you can rewrite the code to make it more parallel. For example, you can insert a Feedback Node ( ) to pass the results from one calculation to the next on the following iteration of the loop—this also known as pipelining. You can use the pipelining technique to reduce the length of each run through the SCTL by breaking up the code among multiple iterations of the SCTL.

The **Quotient and Remainder** function is another one that cannot be used in a SCTL. If you need to divide a number by an integer value, you can use the **Scale by Power of 2** function instead. This function lets you multiply or divide by powers of two, i.e., 2, 4, 8, 16, 32, etc. For a fixed-point result, you can use the Fixed-Point Math Library for LabVIEW FPGA. The fixed-point divide subVI and configuration panel is shown in Fig. 2.46, including the Execution Mode control which enables the function to be used within a SCTL.

The Fixed-Point math Library contains LabVIEW FPGA IP blocks that implement a variety of elementary and transcendental math functions. These functions use the Fixed-point data type introduced in LabVIEW 8.5 extending the current offering of functions to include Divide, Sine, Cosine, and many more important math operations. All functions are verified for usage inside and outside a SCTL as well as in Windows and FPGA simulation on the Development computer. The toolkit comes with help documentation that includes details for each function to learn more about individual usage. To download the free toolkit, follow the link below [9].

If you are trying to make a subVI for use inside of a SCTL, you can use a feedback node to hold state information in the subVI. This eliminates the need to use a while loop inside of a SCTL. The LabVIEW FPGA example below calculates one of the differential equations for a DC motor using functions from the

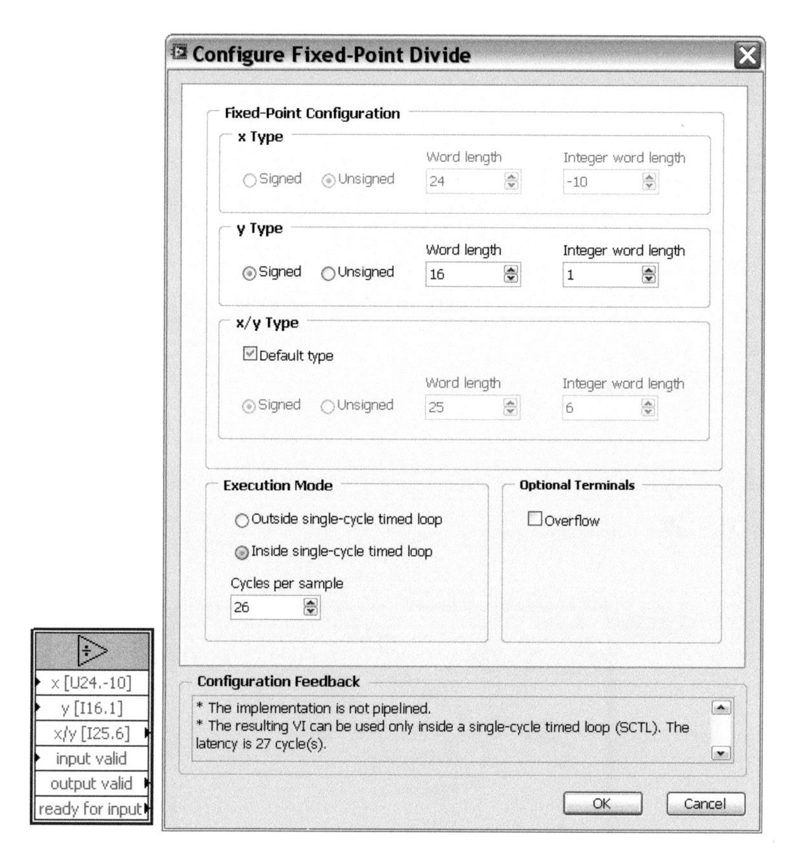

Fig. 2.46 Fixed-point divide

fixed-point math library (see Fig. 2.47). After each fixed-point math function, a feedback node is used to pipeline the result and thereby pass the value from one iteration to the next. In the upper right corner, a **Tick Count** function is also used in combination with a feedback node to calculate the loop rate of the subVI execution.

In Fig. 2.48 you can see the top-level SCTL in the FPGA application, which calls the motor simulation subVI. Since the subVI is nested within a SCTL, the **Loop Rate (Ticks)** value returned is always equal to 1. However, due to pipelining there is a six-tick latency from the **voltage (V)** input to the **i (A)** current output of the subVI.

In addition to pipelining, you can use a State machine within the SCTL to better organize your code and run through a sequence of steps. The basic component of the State machine is a Case structure with each containing one state and using a shift register to determine the next state after each iteration of the loop. Of course each state must be able to run in one clock cycle if the subVI is to be placed in a SCTL. In addition, you can use shift registers and a counter value to implement the functionality of a **For Loop** or add a specific number of Wait states to your program execution.

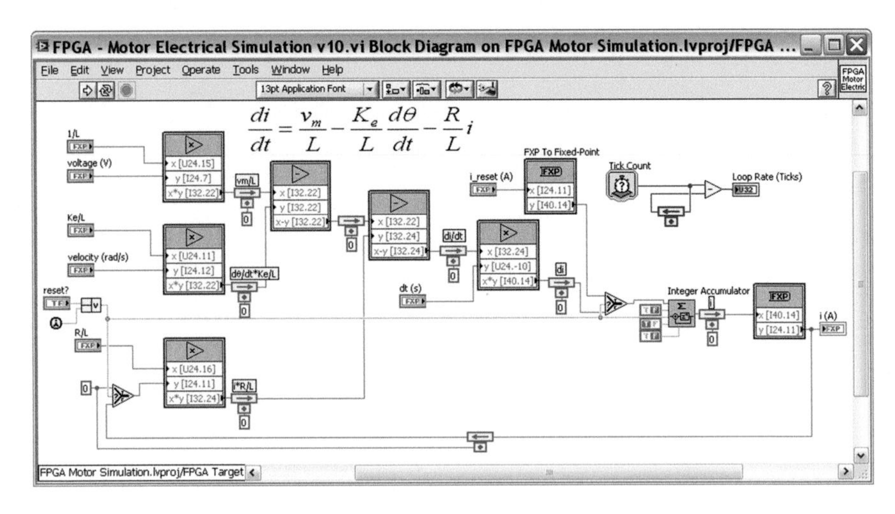

Fig. 2.47 DC motor deferential equation

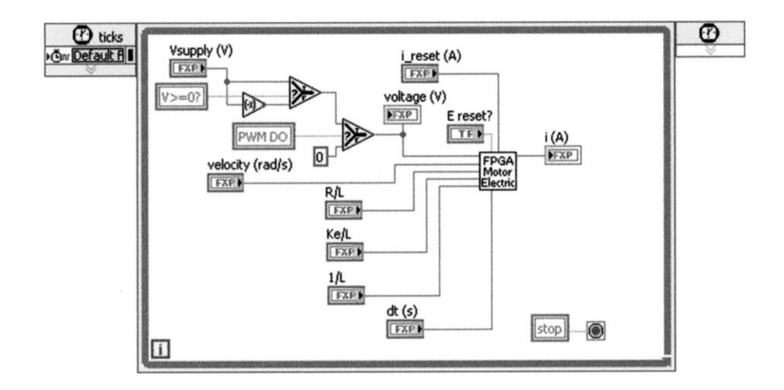

Fig. 2.48 SCTL in the FPGA application

Note: Adding a loop timer or wait function will cause the code to execute slower than one tick, and therefore cannot be used within a SCTL. Analog input and analog output functions also take more than one clock tick to execute and cannot be used in a SCTL. However, you can put them a normal while loops and use local variables to share data with the SCTLs.

2.4.3 Creating Counters and Timers

If you need to trigger an event after a period of time, use the Tick Count function to measure elapsed time as shown in Fig. 2.49. Do not use the iteration terminal that is built into while loops and SCTLs because it will eventually saturate at its maximum value. This happens after 2,147,483,647 iterations of the loop. At a 40 MHz clock

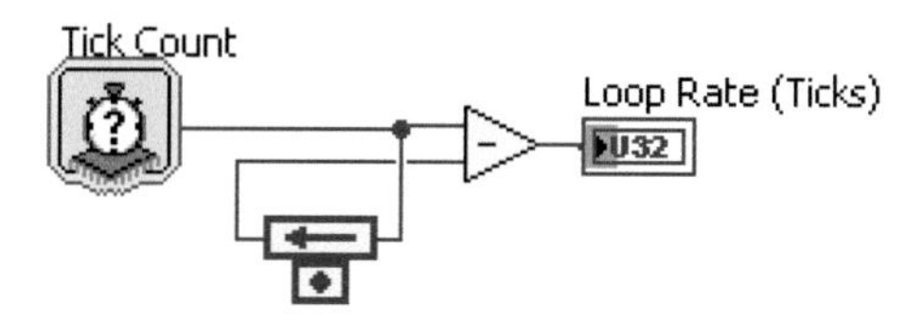

Fig. 2.49 Tick count function to measure elapsed time

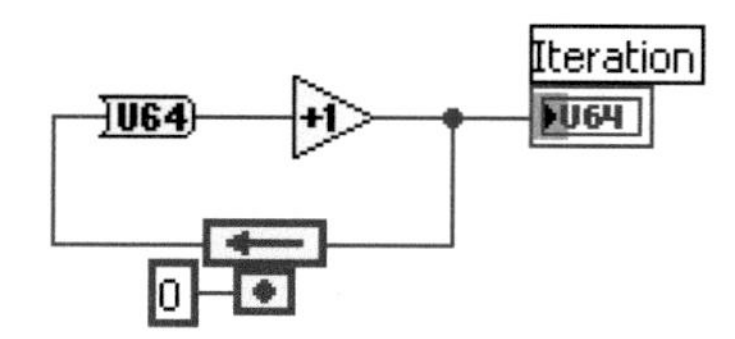

Fig. 2.50 An iteration counter

rate, this takes only 53.687 s. Instead, make your own counter using an unsigned integer and a feedback node. The tick count function is to provide time based on the 40 MHz FPGA clock.

By using an unsigned integer for the counter value, elapsed time calculations will still be correct when the counter rolls over. This is because if you subtract one count value from another using unsigned integers, you still get the correct answer even if the counters overflows.

Another common type of counter is an iteration counter that measures the number of times a loop has executed (see Fig. 2.50). Unsigned integers are typically preferred for iteration counters because they give the largest range before rolling over. The unsigned 64-bit integer data type we are using for the counter provides a huge counting range—equivalent to about 18 billion–billion. Even with the FPGA clock running at 40 MHz, this counter will not overflow for more than 14,000 years.

Now let us talk about another technique that will help you create well-written and efficient LabVIEW FPGA code.

2.4.4 Write Your FPGA Code as Modular, Reusable SubVIs

The next major development technique we will suggest is modular development—break your application into independent functions that can each be individually specified, designed, and tested. It seems like a simple concept, but for FPGA development it can have some especially nice benefits. Here is a simple example—a function designed to measure the rate of the loop in which it is placed and count the number of time it executes. Inside the loop we have a **Tick Count** function that reads the current FPGA clock and subtracts it from the previous value, which is stored in a shift register. In addition, we have a 64-bit counter that increments each

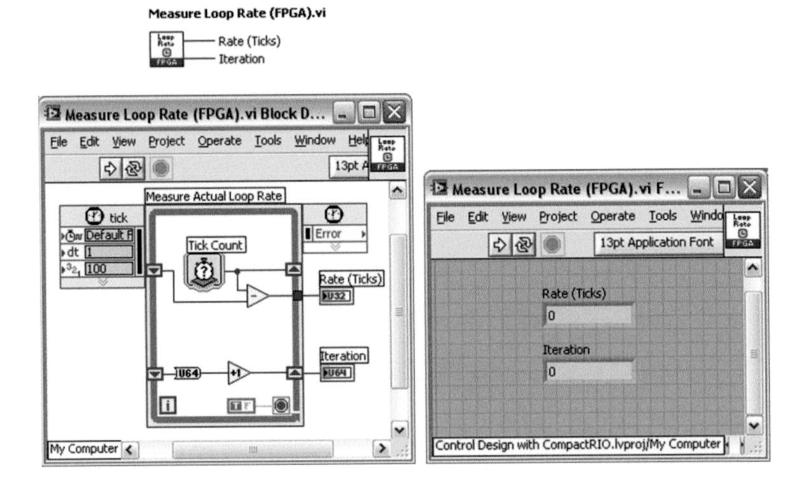

Fig. 2.51 FPGA subVI

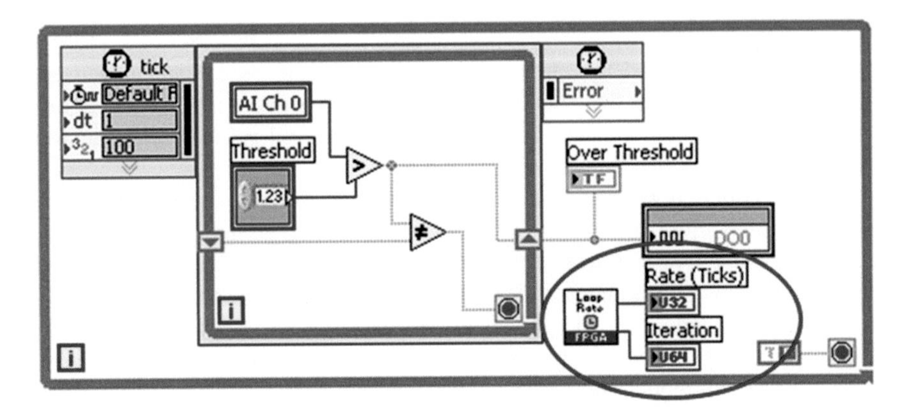

Fig. 2.52 The subVI placed inside another loop

time the function is called. This function uses a SCTL so it only takes a single 25 ns clock tick to execute—therefore, this subVI is designed to be placed inside of a normal while loop without affecting its execution speed (see Fig. 2.51).

Here is the front panel. The indicators have been assigned to the two right terminals of the subVI so data can be passed to the upper level LabVIEW FPGA application in which it is placed.

Fig. 2.52 shows an application example where the function is used. The subVI is placed inside another loop to measure the execution rate of the top-level code.

Some of the top benefits of writing the code this way are presented in Table 2.3.

Writing modular code is almost always a good idea, but when you are designing FPGA logic it has extra advantages.

Table 2.3 Top benefits of writing the code

Benefit	Explanation
Easier to debug and troubleshoot	Code can be tested on Windows before compiling
Easier to document and track changes	Help information can be included in the VI documentation
Creates a cleaner, more easily understood top-level diagram	Code is more intuitive to other programmers

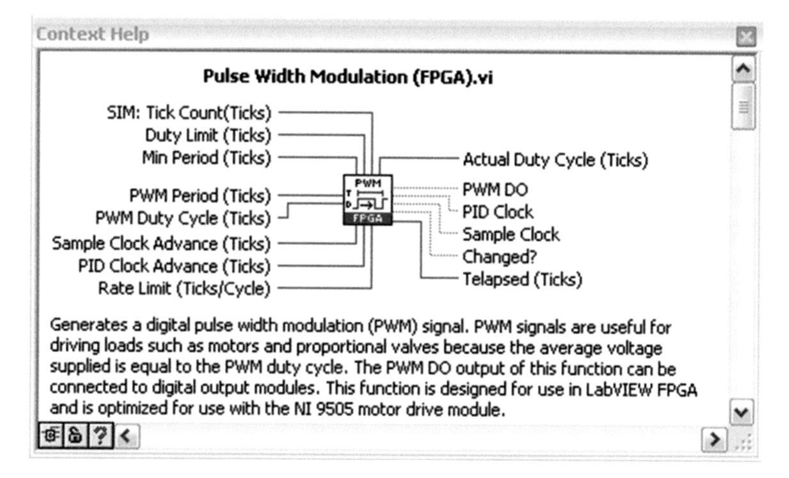

Fig. 2.53 PWM example

First, the code is easier to debug and troubleshoot. One big benefit is that the subVI can be tested on Windows before you compile it for the FPGA. We will show some examples of that later.

Second, it is easier to document and track changes because the code is modular and you can include help information in the VI documentation.

Third, the intended functionality of the code is typically cleaner, easier to understand, and more reusable. The options you want to offer the programmer are typically made available as terminals on the subVI. Most often the user will not need to modify the underlying code—they can just use the parameters you provide, such as in this **Pulse Width Modulation (FPGA)** example (see Fig. 2.53).

Now let us take a look at some tips for how to create modular, reusable subVIs for LabVIEW FPGA.

2.4.5 *Separate Logic from I/O*

The first tip is to keep I/O nodes out of your subVIs. This makes them more modular and portable and makes the top-level diagram more readable. Particularly for control applications, it is nice to have all of the I/O operations made clearly visible when viewing the top-level diagram for the application, like we have shown here in this PWM loop written for the NI 9505 motor drive module (see Fig. 2.54).

Rather than embedding the I/O node into the subVI, a terminal is used to pass the data from the subVI to the top-level diagram. This makes the FPGA code easier to debug, since the subVI can be tested individually in Windows using simulated I/O. This will be explained in more detail in a subsequent section (see Fig. 2.55).

Taking this approach also tends to reduce unnecessary I/O node instances that might otherwise be included multiple times in the subVI, resulting in unnecessary gate usage due to the need for the compiler to add the extra arbitration logic necessary to handle multiple callers accessing the same shared resource.

Finally, this approach makes the top-level application more readable—all I/O read and write operations are explicitly shown on the top-level diagram and not hidden from view.

Often when you are writing function blocks like this, the subVI will need some local memory capability so it can hold state values, such as elapsed time, and pass that information from one iteration to the next.

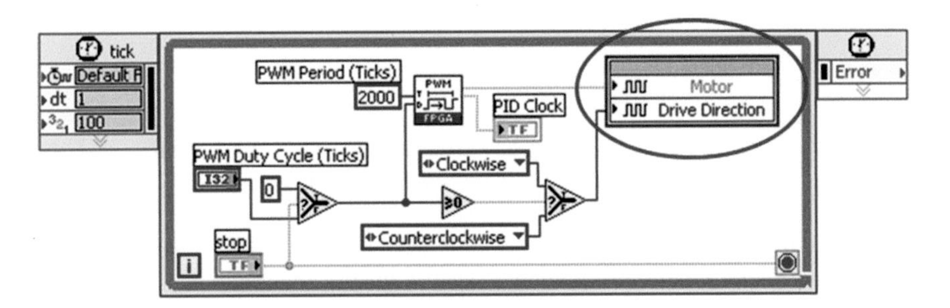

Fig. 2.54 PWM example for the NI 9505

Pulse Width Modulation (FPGA).vi

Duty Limit (Ticks)
Min Period (Ticks)
PWM Period (Ticks)
PWM Duty Cycle (Ticks)
Sample Clock Advance (Ticks)
PID Clock Advance (Ticks)
Rate Limit (Ticks/Cycle)
PWM
FPGA
Actual Duty Cycle (Ticks)
PWM DO
PID Clock
Sample Clock
Changed?
Telapsed (Ticks)

Fig. 2.55 PWM SubVI

2.4.6 Holding State Values in a Function Block

In the example below, you can see that we have added shift register nodes to our loop that enable us to pass information from one iteration of the loop to the next. The iteration counter increments each time the function block is called.

Notice that the **Loop Condition** terminal has a constant wired to it that causes the loop to run for only one iteration each time it is called. We are not really looping in this case—we are simply using the SCTL to optimize the code and to hold the state values using shift registers (see Fig. 2.56).

Note: It is important to note that the shift register must be un-initialized for the subVI to hold state this way. On first call the shift register value is the default value for the data type—for integers that is 0, for Booleans that is False. If you need to initialize the value to something else, use a **First Call?** function and a **Select** function to initialize the value.

You may be wondering how to create a modular function block that works inside a SCTL, since you are not allowed to nest one SCTL within another.

To do this, use feedback nodes to accomplish the same task, as shown below. The main benefit of this approach is that the feedback nodes can be easily initialized, and the subVI could now be placed within a top-level SCTL because it contains no nested loop structure (see Fig. 2.57).

A third option is the use of VI-scoped memory (see Fig. 2.58). This is a block of memory that can be used locally by the subVI and does not have to be manually

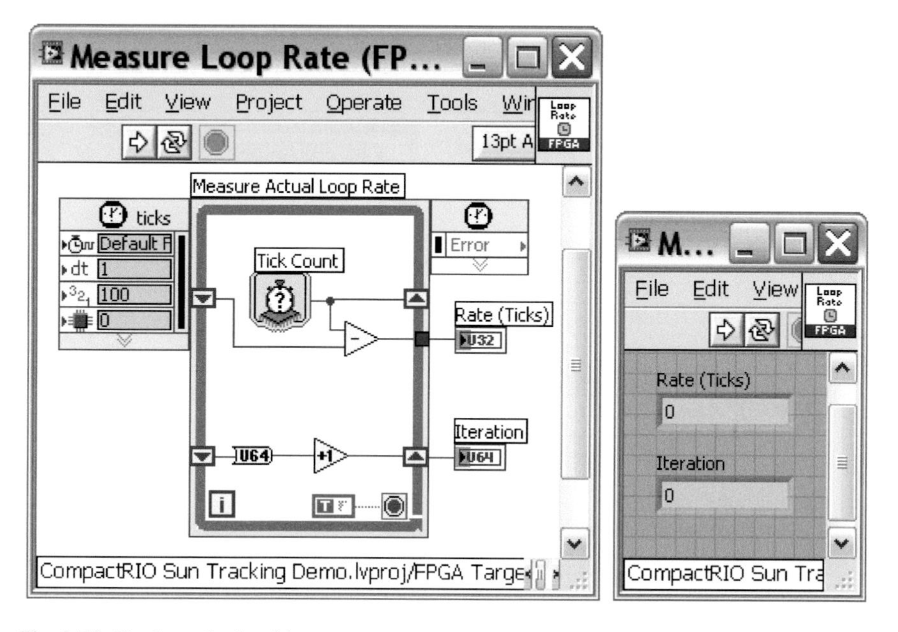

Fig. 2.56 Single-cycle timed loop

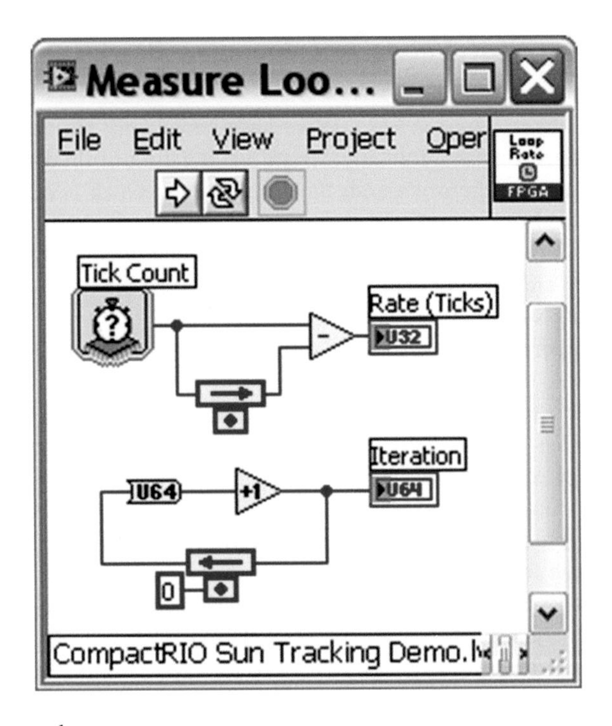

Fig. 2.57 Measure loop

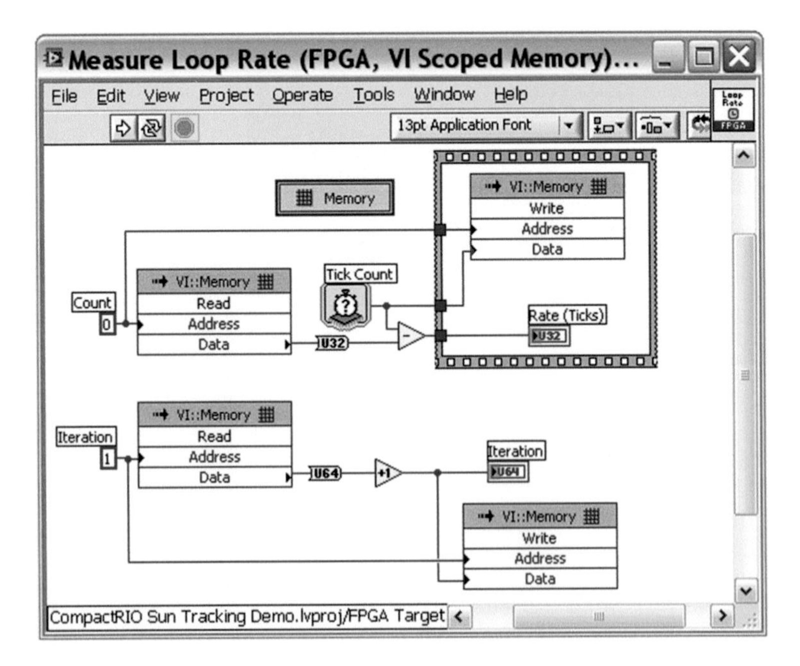

Fig. 2.58 VI memory

added to the project. This makes the code more modular and portable when moving it between projects.

In this simple example, using VI-scoped memory is probably overkill for the application. We only have two memory locations and we are only storing one data point in each memory location. However, VI-scoped memory is a powerful tool for applications that need to store arrays of data. In general, you should always avoid using large front panel arrays as a data storage mechanism—use VI-scoped memory instead.

2.4.7 Run-Time Updateable Look-up Table (LUT)

A common use for local memory in FPGA-based control applications is to store table data, such as the calibration table for a nonlinear sensor, a pre-calculated math formula (such as log or exponential), or an arbitrary waveform that can be replayed by indexing through the table addresses. Below is an FPGA-based look-up table (LUT) configured to store 10,000 fixed-point values and perform linear interpolation between stored values. Because VI-scoped memory is used, the LUT values can be changed while the application is running and without the need to recompile the FPGA (see Figs. 2.59 and 2.60).

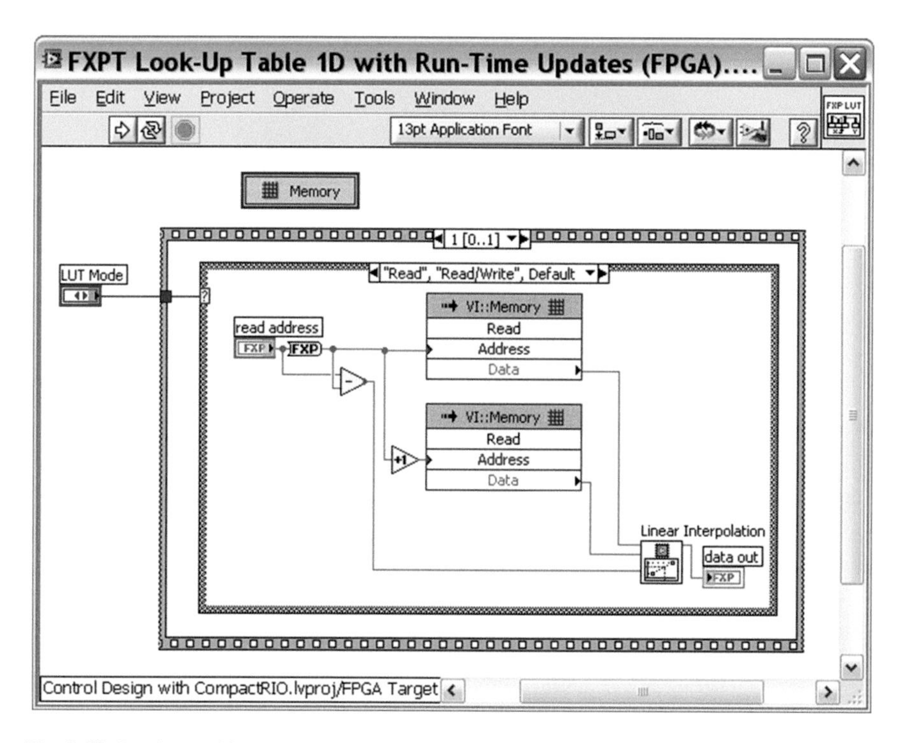

Fig. 2.59 Look-up table

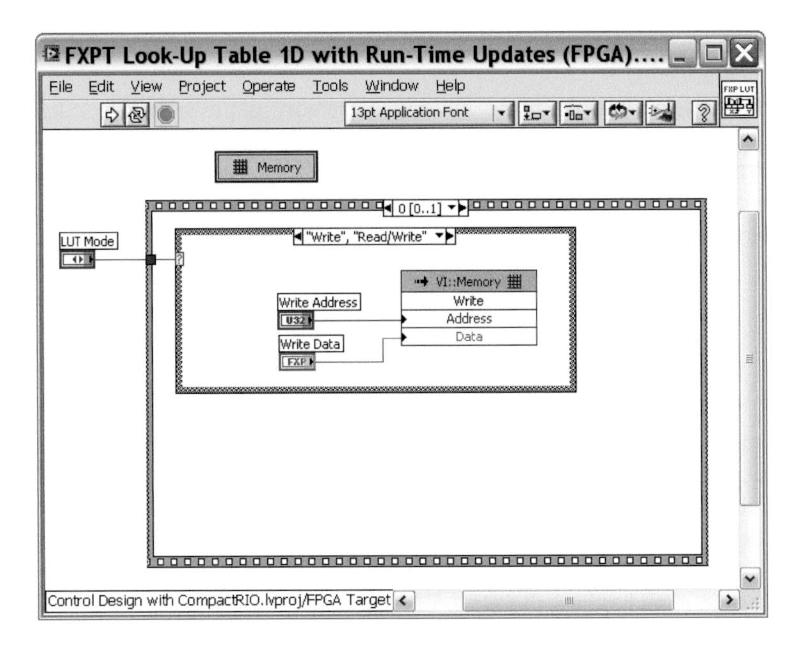

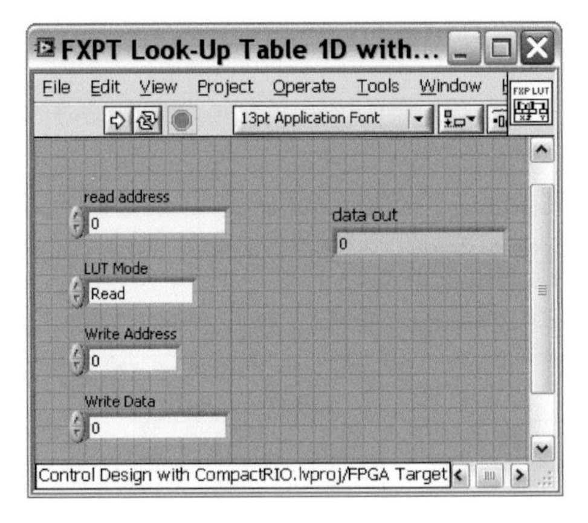

Fig. 2.60 Fix point frontal panel and block diagram

Let us take a look at the configuration pages for the VI-Scoped Memory block in this example. You can configure the depth, data type, and even define initial values for the memory elements (see Fig. 2.61).

Now let us look at another tip for creating modular FPGA subVIs that have to do with the timing of how the code runs.

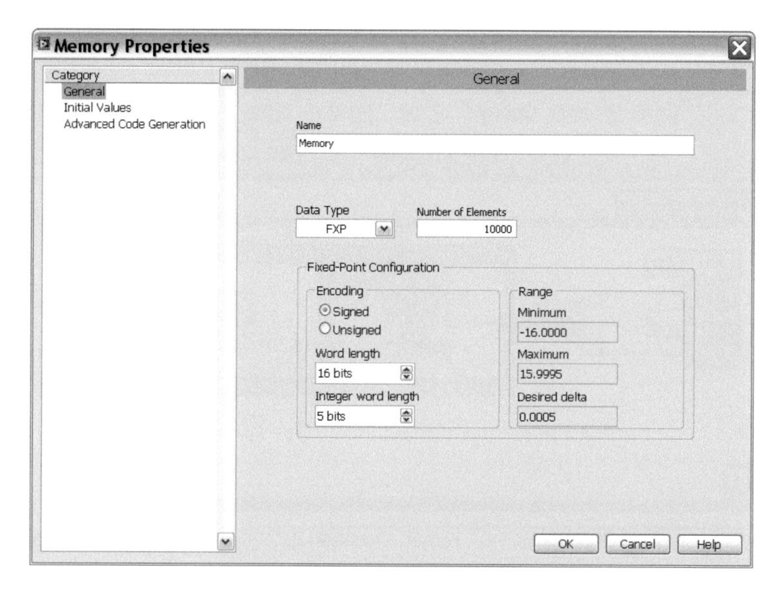

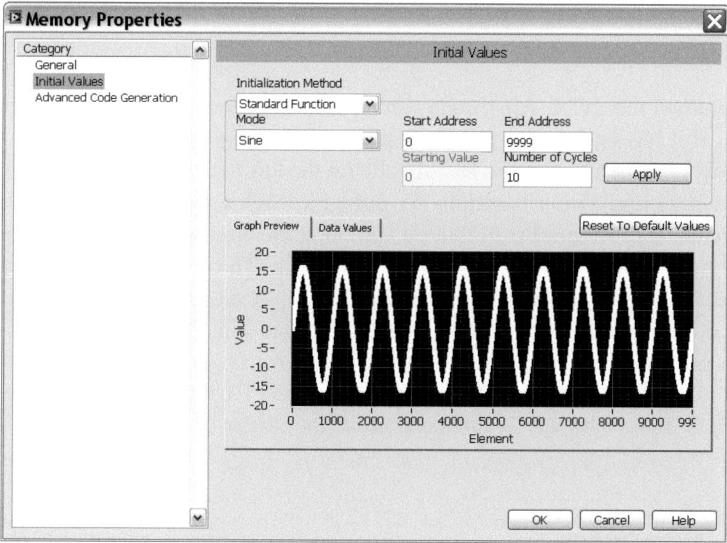

Fig. 2.61 Memory properties

2.4.8 Do not Place Delay Timers in the SubVI

In general, it is a good idea to avoid using Loop Timer or Wait functions within your modular subVIs. If the subVI has no delays, it will execute "as fast as possible" and thereby inherent the timing properties of the calling VI, rather than

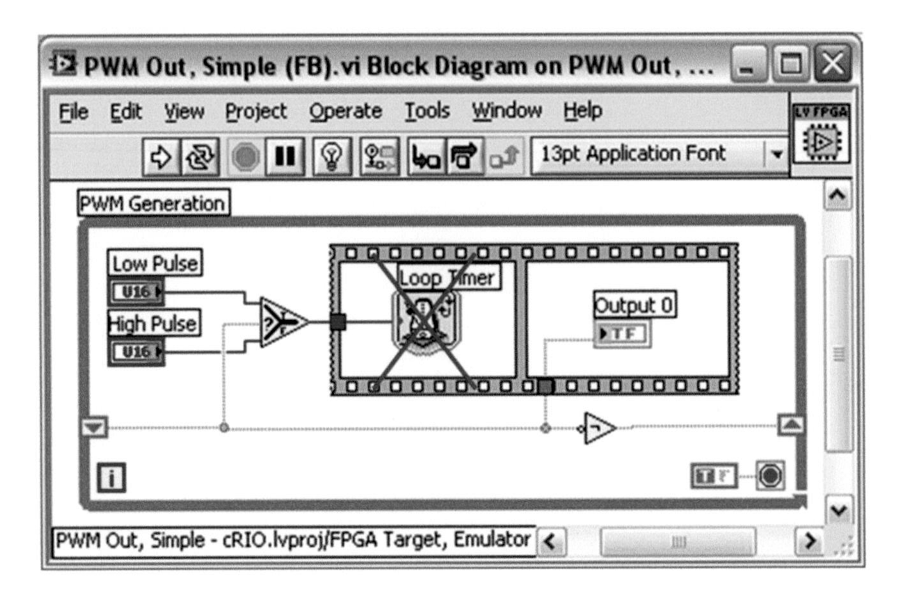

Fig. 2.62 Do not place delay times

slowing down the caller. Also, code can typically be more easily adapted for use in a SCTL if it has no internal functions that cause delays (see Fig. 2.62).

Below we have adapted the PWM code on the left to use a **Tick Count** function rather than a **Loop Timer** function. We use a feedback node to hold an elapsed time count value, and we turn the output on and off at the appropriate times and reset the elapsed time counter at the end of the PWM cycle. The code may look a bit more complicated, but it can be dropped inside of a top-level loop without affecting the overall timing of the loop—it is more portable (see Fig. 2.63).

Now let me share one more tip before we move on to the next topic—how to make the code so that multiple copies of a subVI can be placed in the same application and each copy is independent of the others.

2.4.9 *Reentrancy*

Reentrancy is a setting in the subVI execution properties that enable multiple copies of the function block to be executed in parallel with distinct and separate data storage (see Fig. 2.64).

Figure 2.65 shows an example. In this case our subVI is set to reentrant, meaning all four of these loops will run simultaneously and any internal shift registers, local variables, or VI-scoped memory data will be unique to each instance.

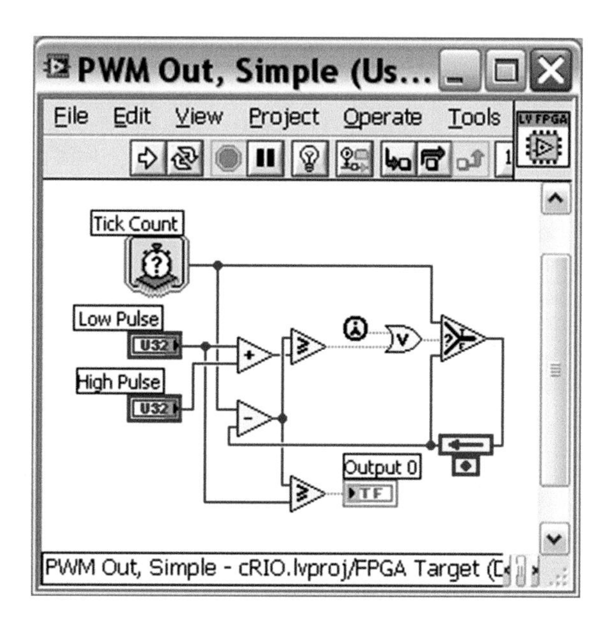

Fig. 2.63 Tick count function

In the case of LabVIEW FPGA, it also means that each copy of the function uses its own FPGA slices—so reentrancy is great for code portability but it does use more gates.

Note: If you are really squeezed for FPGA gates, you can make your function multiplexed rather than reentrant. This is an advanced topic we will not cover here but it basically involves using local memory to store the register values for each of the calling loops, which identify themselves with an integer "ID tag" value. Since

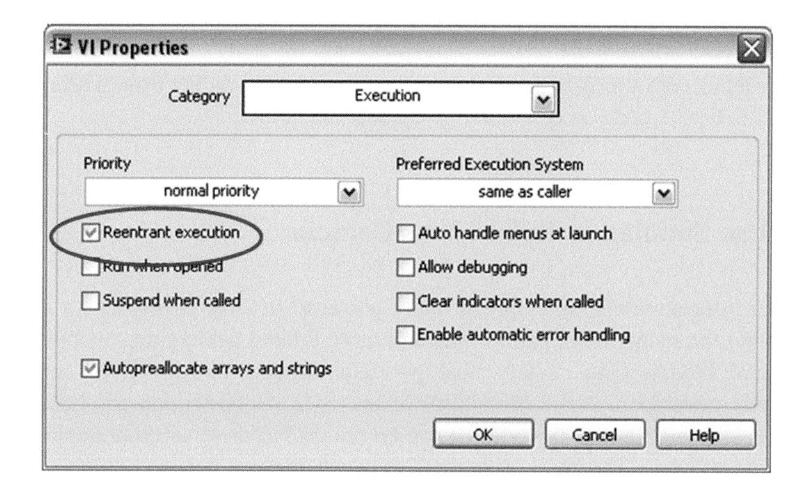

Fig. 2.64 Reentrancy

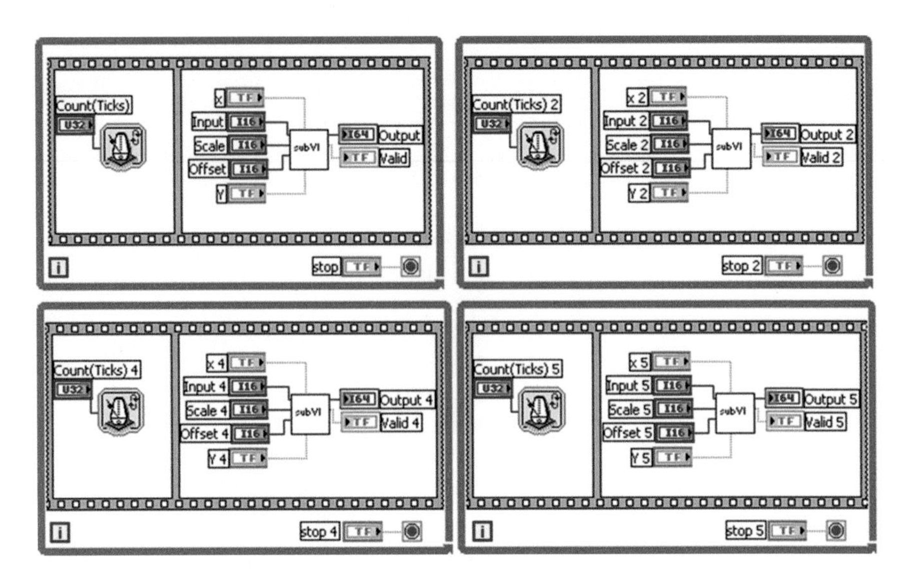

Fig. 2.65 SubVI is set to reentrant

the loops are all using the same underlying FPGA slices (with different memory addresses for the data), each caller will block the other callers resulting in slower execution. However, gate usage is much less since the same hardware SLICE logic is reused. For many control applications where the FPGA is already much faster than the I/O, this is a nice option for saving gates. Several functions on the LabVIEW FPGA palette use multiplexing techniques to enable high channel count operation with minimal FPGA gate usage. These include the **PID**, **Butterworth Filter**, **Notch Filter**, and **Rational Resampler** functions. To see how this works, drop one of these functions onto the block diagram and configure it for multiple channels. Then right-click on the function and select **Convert to SubVI** to reveal the underlying code.

Now let us take a look at a major development benefit you get from writing your LabVIEW FPGA code as described in the sections above.

2.5 Use Simulation Before You Compile

This third development technique is really powerful because it provides a way to get around the longer compilation time and more limited debugging capabilities of LabVIEW FPGA. One of the most powerful aspects of LabVIEW code for embedded developers is the portability of the code. Code written for LabVIEW FPGA is still just LabVIEW code—it can be run on Windows or other devices and operating systems. The main difference between these processing targets is the speed at which the code runs and whether they support true parallel processing like

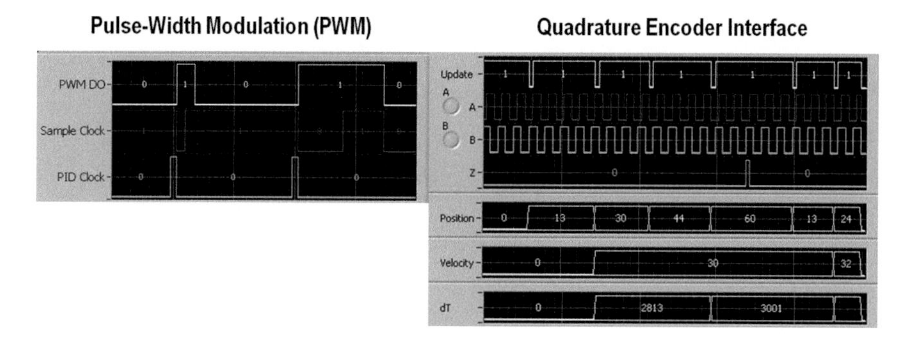

Fig. 2.66 FPGA simulation

an FPGA or simulated parallel processing like a multithreaded operating system for a microprocessor.

LabVIEW FPGA includes the ability to run the entire LabVIEW FPGA application in simulation mode, and this can be done in conjunction with the host processor application for testing purposes with either random data used for FPGA I/O read operations or using a custom VI to generate the simulated I/O signals. This is particularly useful for testing FPGA to host communication including DMA data transfers.

However, the disadvantage of this approach is that the entire FPGA application is simulated. For the development and testing of new LabVIEW functions, it can be advantageous to test the code one function at a time. This section will focus on a capability called functional simulation, which enables a "divide and conquer" approach to debugging which allows each function to be tested individually before compiling to the FPGA. Below are screens from two functional simulation examples running on Windows that were used for testing and debugging purposes (see Fig. 2.66).

The example below shows the front panel and block diagram of a test application used to debug a LabVIEW FPGA subVI for PWM. The test application is located in the **My Computer** section of the LabVIEW project, and when it is opened it runs in Windows (see Fig. 2.67).

2.5.1 Providing Tick Count Values for Simulation

The Conditional Disable Structure in LabVIEW lets you modify what underlying code is used when the subVI is compiled for different processing targets. In this case, I have got a Tick Count function that is executed when the code is compiled for the FPGA and a front panel control that is executed when the code is executed on Windows. This lets me use a "simulated" tick count value when I am testing the

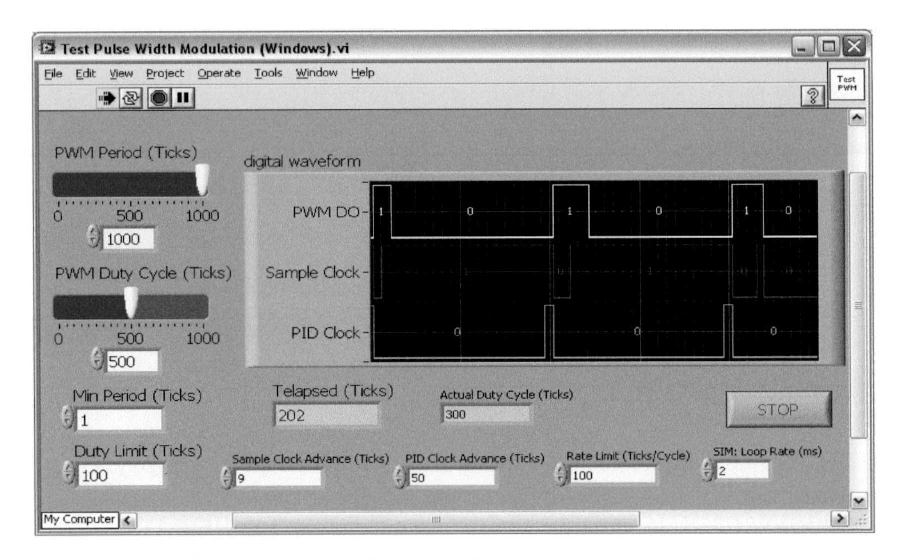

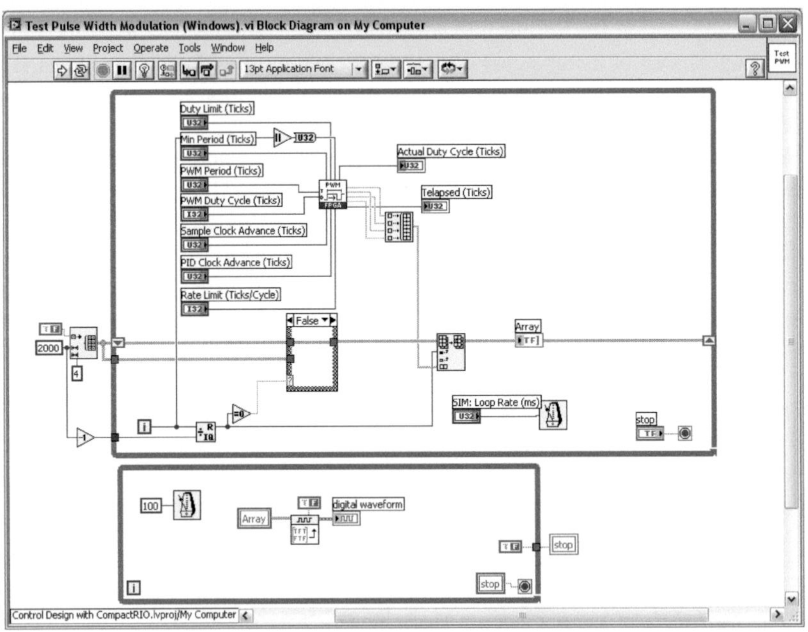

Fig. 2.67 Debugging a LabVIEW FPGA subVI for pulse width modulation

code in Windows, providing the ability to create both bit accurate and cycle accurate simulations (see Fig. 2.68).

This technique is used in the PWM test example above—when the subVI is executed in Windows a simulated FPGA clock is passed to the subVI using the Iteration terminal of the top-level while loop.

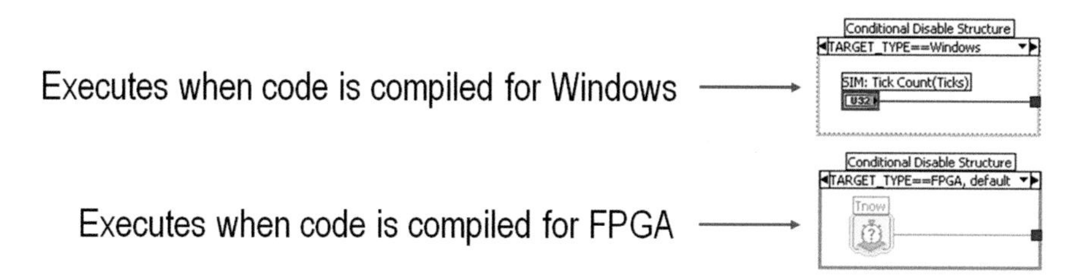

Fig. 2.68 FPGA code compiled by Windows and FPGA

As you have seen, functional simulation lets you test, iterate and be confident in your FPGA logic before you compile. It also enables you to use the full LabVIEW debugging toolset while the code is running, and you can create "test patterns" that enable you to verify the code under a variety of conditions that otherwise might be hard to test. Here are some of the top benefits of using simulation as a step in your development process.

- Quickly iterate and add features
- Be confident in your LabVIEW FPGA code before you compile
- Use full LabVIEW debugging capabilities (probes, highlight execution, etc.)
- Verify the code under a variety of conditions

Now let us take simulation a step farther and create a simulation that accurately mimics the dynamic closed-loop behavior of the physical system within which our LabVIEW FPGA code will be connected.

2.5.2 *Test the LabVIEW FPGA Code Using the LabVIEW Control Design & Simulation Module*

The LabVIEW Control Design & Simulation Module (CD&Sim) includes state-of-the-art technology for simulating mechatronic systems like the DC motor we will be controlling with our LabVIEW FPGA application. Figure 2.69 shows the theoretical model equations for a brushed DC motor driven by a PWM chopper circuit and connected to a simple inertial load with viscous friction.

This is implemented using a LabVIEW CD&Sim subsystem containing formula node. The two differential equations shown above are entered into the formula nodes in text format as shown below. Integrator functions ($\frac{1}{s}$) are used to convert from higher order derivatives, such as from acceleration to velocity and from velocity to position (see Fig. 2.70).

The Brushed DC Motor.vi subsystem is placed within a top-level simulation loop and connected to the LabVIEW FPGA function to simulate the pulsed voltage signal used to drive the motor. The result is a high fidelity closed-loop simulation of

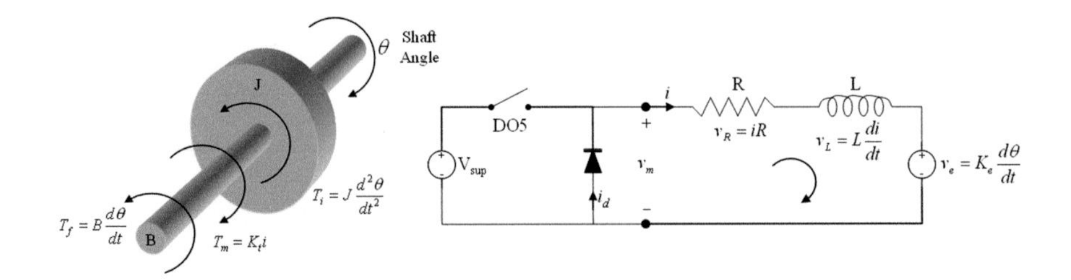

Sum of Forces

$$\frac{d^2\theta}{dt^2} = \frac{K_t}{J}i - \frac{B}{J}\frac{d\theta}{dt}$$

Kirchhoff's Voltage Law

$$\frac{di}{dt} = -\frac{R}{L}i - \frac{K_e}{L}\frac{d\theta}{dt} + \frac{v_m}{L}$$

Fig. 2.69 DC drive motor model

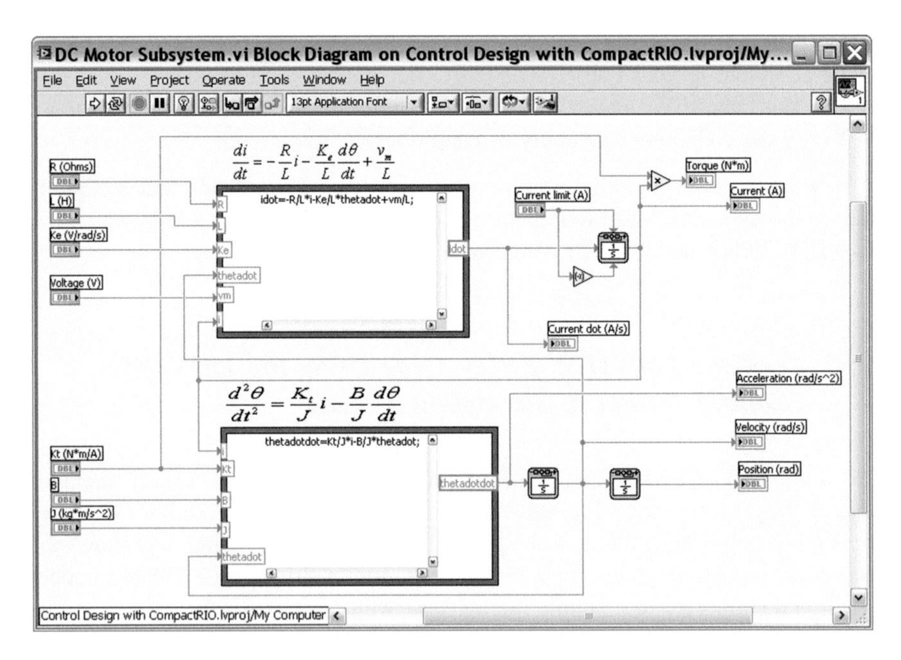

Fig. 2.70 DC drive model

how the LabVIEW FPGA code will behave when connected to the real-world electromechanical system (see Fig. 2.71).

The simulation results have been validated against actual measurements from the deployed LabVIEW FPGA application controlling a motor using the NI 9505 motor drive module, which showed a nearly identical match between the simulated and measured waveforms.

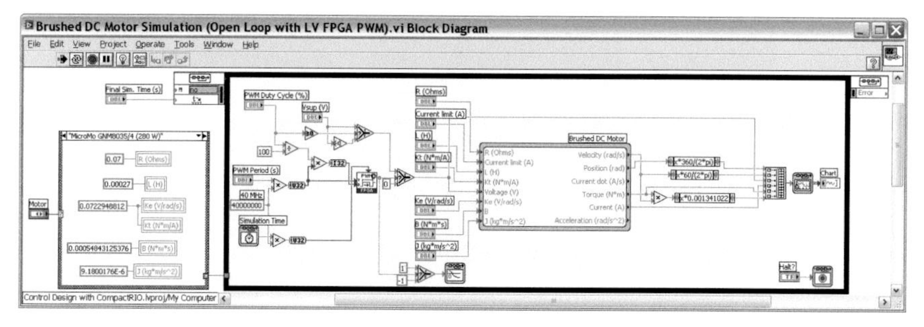

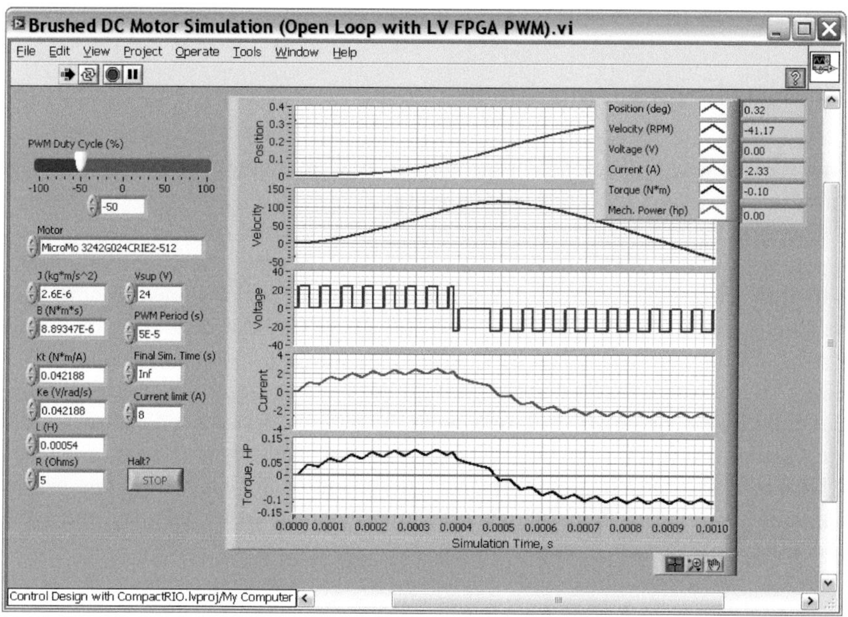

Fig. 2.71 Brushless motor drive

This approach lets you take code validation way beyond basic functional validation. Think of this like a virtual machine emulator that lets you anticipate how your code will perform in the real world. You can use simulation to help make design decisions, evaluate performance, select components, and test worst-case conditions. You can even tune the PID control loops for your control system in simulation and see how well that tuning works with different motors and load conditions. Simulation can also help you select the right physical components for your system, such as picking the right motor to meet your performance requirements.

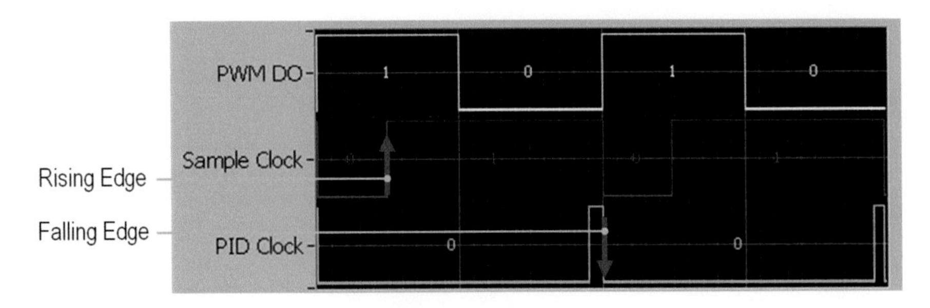

Fig. 2.72 Trigger on either the rising or falling edge

2.6 Synchronize Your Loops

Now for our fourth development technique—how to control the timing and synchronization of your LabVIEW FPGA code.

For most control applications, the timing of when the code executes is very important to the performance and reliability of the system. Fortunately, LabVIEW FPGA gives you both unprecedented speed and complete control over the timing of the code. Unlike a processor, an FPGA executes code in a truly parallel fashion rather than only being able to execute one instruction at a time. That makes programming easier because you do not have to worry about setting priorities and sharing the processor time among the different tasks. Each control loop is like a custom designed processor that is completely dedicated to its task. The result is high reliability and high-performance code. One of the benefits of this performance is that control loops are typically more stable, easier to tune, and more responsive to disturbance when they run at a fast rate.

In this motor control example, we have two different clock signals—a Sample Clock and a PID Clock. These are Boolean signals we generate in the application to provide synchronization among the loops. We can trigger on either the rising or falling edge of these clock signals (see Fig. 2.72).

Now let us take a look at the LabVIEW FPGA code used to monitor these signals and trigger on either the rising or falling edge.

Typically triggering a loop based on a Boolean clock signal works like this—first wait for the rising or falling edge to occur, and then execute the LabVIEW FPGA code that you want to run when the trigger condition occurs. A sequence structure is often used where the first frame of the sequence is used to wait for the trigger, and the second frame is used to execute the triggered code, as shown below.

Rising Edge Trigger: In this case we are looking for the trigger signal to transition from False (or 0) to True (or 1). This is done by holding the value in a shift register and using the **Greater Than?** Function (see Fig. 2.73). (Note: A True constant is wired to the iteration terminal to initialize the value and avoid an early trigger on the first iteration.)

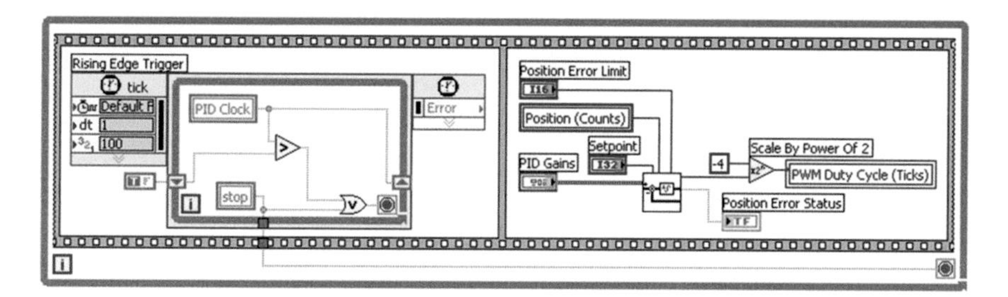

Fig. 2.73 Rising edge trigger

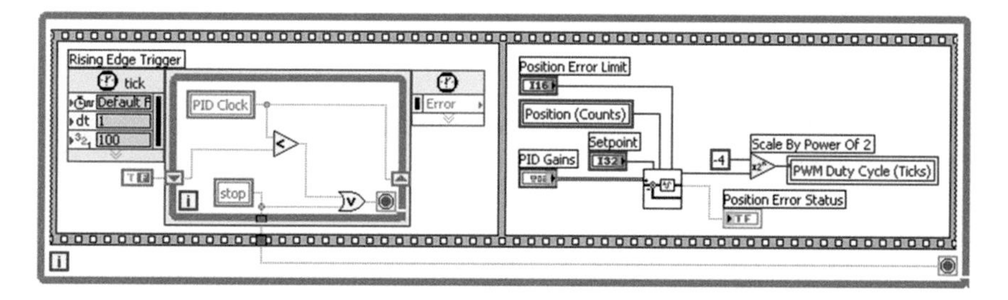

Fig. 2.74 Falling edge trigger

Falling Edge Trigger: In this case we use a **Less Than?** function to detect the transition from True (or 1) to False (or 0) (see Fig. 2.74). (Note: A False constant is wired to the iteration terminal to initialize the value.)

Analog Level Trigger: Here we use a **Greater Than?** function to detect when the analog signal is greater than our analog threshold level, and then use the Boolean output of the function as our trigger signal (see Fig. 2.75). This case actually a rising **or** falling edge detector since we are using the Not Equal? function to detect any transition.

Now let us take a look at another common triggering use case—this is where we want to latch the value of a signal when a trigger event occurs.

2.6.1 Latching Values

In this case we use a rising edge trigger to latch the **Analog Input** (see Fig. 2.76) value from another loop into the **Latched Analog Input** register. This value is held constant until the next trigger event occurs. In this example, the actual analog input operation is occurring in another loop and we are using a local variable for

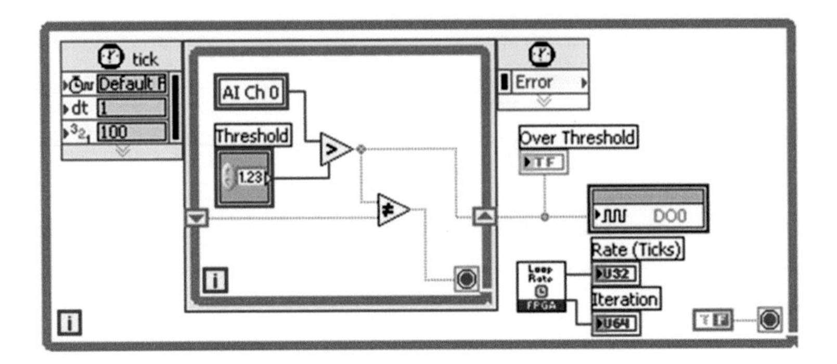

Fig. 2.75 Analog level trigger

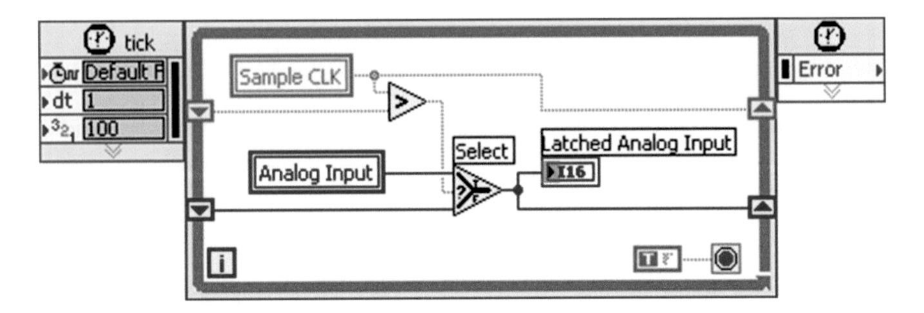

Fig. 2.76 Latched analog input

communication between the loops. (Note: Local variables are a good way to share data between asynchronous loops in LabVIEW FPGA.)

2.6.2 Application Example

Below is an example (see Fig. 2.77) that shows these triggering and latching techniques in practice. LabVIEW FPGA offers true parallel execution. In this case we have three independent loops. This is like having three custom designed processors running at the same time within the chip. Each loop is completely dedicated to its own task—resulting in the highest level of reliability. It also makes programming control in an FPGA easier to architect—unlike with a processor, you do not have to worry your existing code slowing down when you add new code.

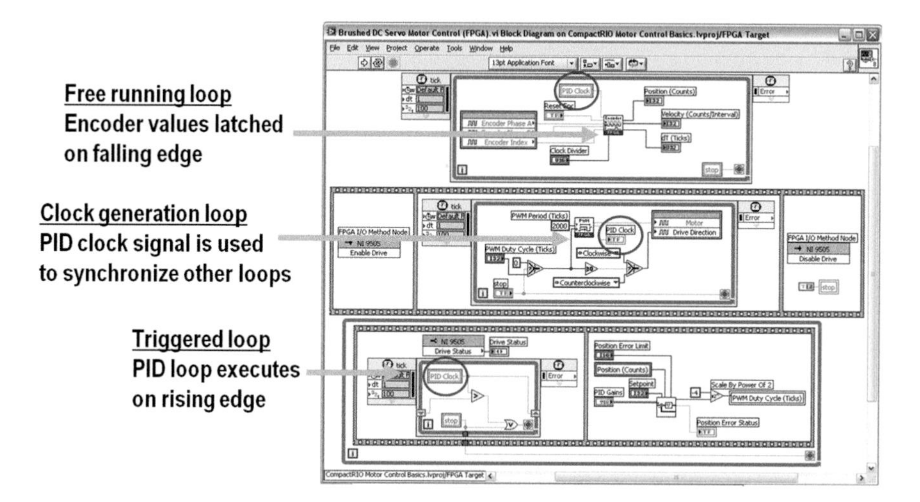

Fig. 2.77 Triggering and latching techniques

Observations:

- One loop is used to generate synchronization clocks used by other loops.
- The encoder function needs to run at full speed to avoid missing any digital pulses. This function runs at 40 MHz but latches the Position (Counts) and Velocity (Counts/Interval) signals to synchronize the data with the other loops.
- The PID function needs to run at a specific speed (20 kHz or 2000 ticks) and avoid any jitter in its timing. This is because the integral and derivate gains depend on the time interval, Ts. If the time interval was varying, or if the same old value was passed multiple times into the function, the integral and derivative gains would be incorrect.
- In the bottom loop, you can see that the execution is triggered by the rising edge of the PID clock signal. We read a local variable for the signal in this SCTL, and exit the loop when a rising edge is detected. Then we execute the 32-bit PID algorithm that is included with the NI SoftMotion Development module. This reads the commanded position, compares it to the position measured by the encoder, and then generates a command for the PWM loop. In this case, we are using a **Scale by Power of 2** function to divide the PID output signal by 2^−4, which is equivalent to dividing by 16. This scales the value to the ±2000 ticks range needed by the PWM function. A value of 1000 ticks is equal to a 50 % duty cycle since the PWM period is 2000 ticks.
- Note that the upper two loops are running at a 40 MHz loop rate, where the lower loop is triggered to run at a 20 kHz loop rate by the PWM clock signal. (When triggered, the SoftMotion PID function takes 36 ticks to execute.)

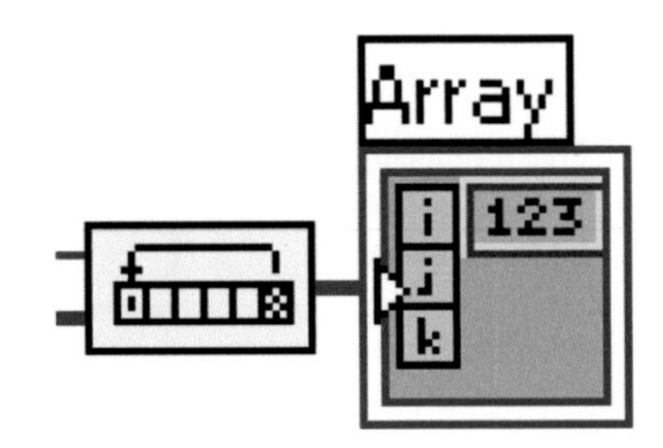

Fig. 2.78 Rotate 1D array function

2.7 Technique 5: Avoid "Gate Hogs"

Now that you understand four key techniques that are useful for developing LabVIEW FPGA code, let us talk about one last technique—how to avoid "gate hogs." These are often "innocent looking" code that eats up lots of your FPGA gates (also known as slices). Here are three of the most common offenders.

Large Arrays or Clusters: Creating a large array or cluster with a front panel indicator or control is one of the most common programming mistakes that eat up lots of FPGA gates. If you do not need a front panel indicator for communication with the host processor, then do not create one. If you need to transfer more than a dozen or so array elements, use DMA instead as a way to pass the data. Also, avoid using array manipulation functions like this **Rotate 1D Array** function whenever possible (see Fig. 2.78).

Quotient and Remainder: This function does integer division. (The quotient output, **floor(x/y)**, is **x** divided by **y**, rounded down to the closest integer. The remainder output, **x-y*floor(x/y)**, is whatever is left over. For example, 23 divide by 5 gives a quotient of 4 and a remainder of 3.) This function is gate intensive and takes multiple clock cycles to execute so it cannot be used in a SCTL. Be sure to wire up the minimum data type needed to the terminals when using this function and use constants rather than controls when possible (see Fig. 2.79).

Scale By Power of 2: If the **n** terminal is positive, this function multiplies the **x** input by 2 to the power of n (**2^n**). If **n** is negative, the function divides by 2^**n**. For example, setting n to **+4** would multiply by **16**, while setting it to **−4** would divide by sixteen. This function is much more efficient than the **Quotient and Remainder** function. However, use a constant of the minimum data size needed for the **n** terminal whenever possible (see Fig. 2.80).

Note: DMA is a better way to send an array of data to the host than creating a front panel indicator for the array and using the **FPGA Read/Write** method. Arrays

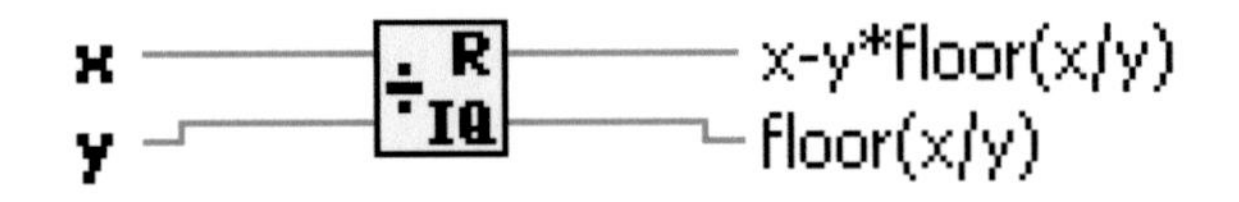

Fig. 2.79 Quotient and remainder

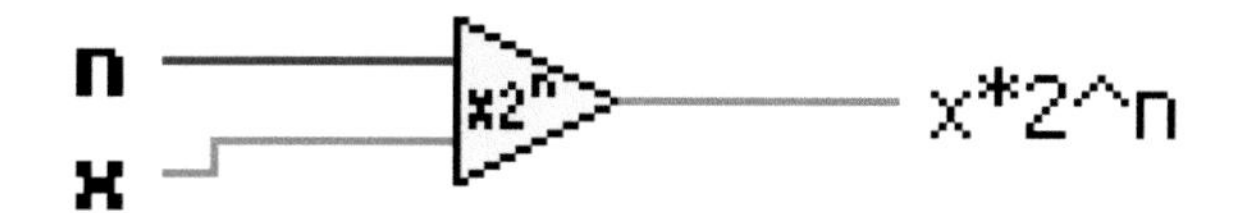

Fig. 2.80 Scale by power of 2

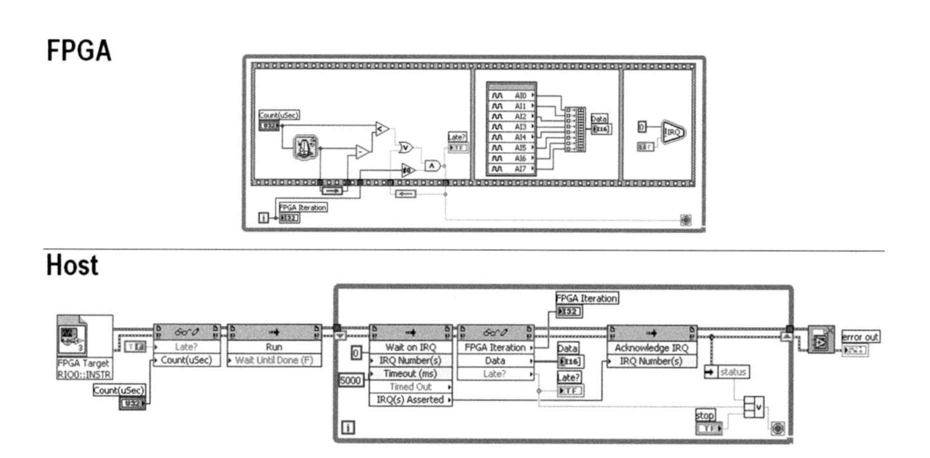

Fig. 2.81 Optimized FPGA code

are useful for collecting a set of simultaneously sampled data to be fed into a DMA buffer for transfer to the host computer. It is okay to use an array to collect the data points together for indexing into the DMA Write function as long as you do not create a front panel indicator for the array. Using auto-indexing on the for loop used to write the data into the DMA, buffer is fine as long as you do not create a front panel indicator for the array because the compiler does a good job of optimizing arrays passed into **For Loops** for indexing purposes.

2.7.1 Avoid Front Panel Arrays for Data Transfer

When optimizing your code for the amount of space it uses on the FPGA, you should consider the front panel controls and indicators you are using. Each front panel object and the data it represents take up a significant portion of the FPGA space. By reducing the number of these objects and reducing the size of any arrays used on the front panel, you can significantly reduce the FPGA space required by the VI (see Fig. 2.81).

Instead of creating large arrays to store data and transfer it to the host application (shown above), use DMA to transfer an array of analog input samples to the host processor as shown in Fig. 2.82.

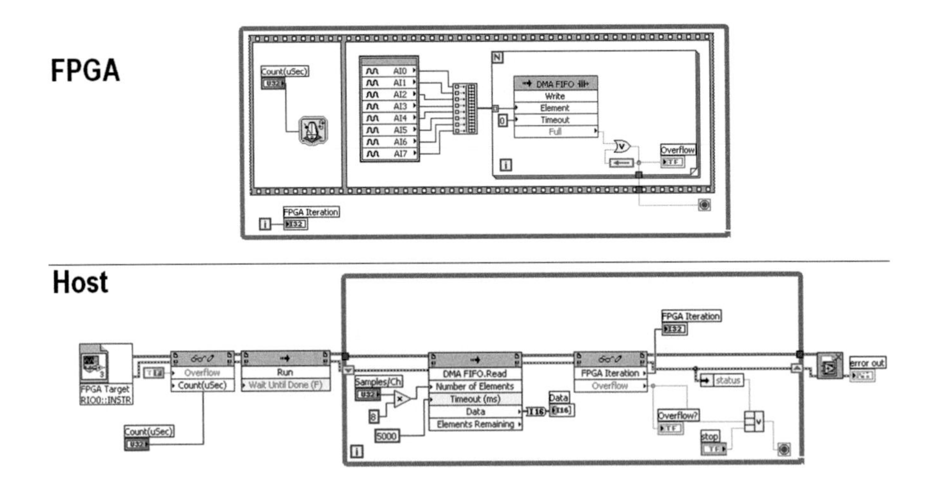

Fig. 2.82 DMA FIFO

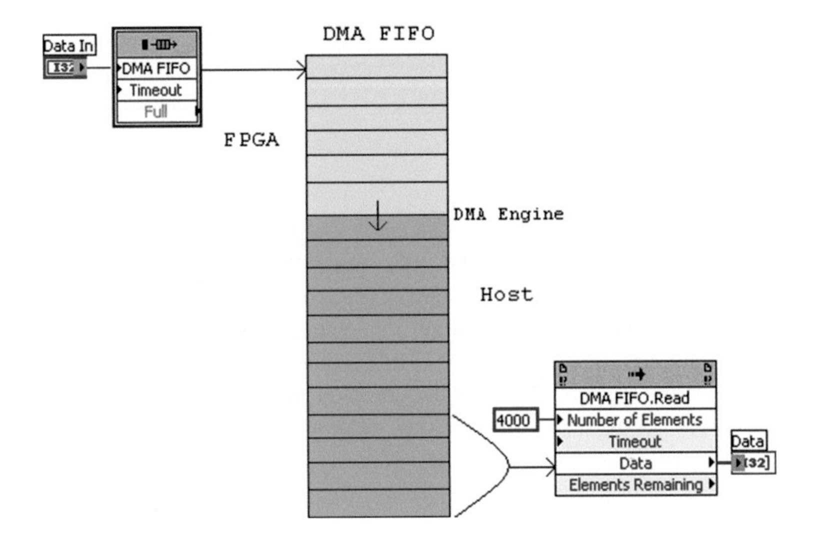

Fig. 2.83 MDA FIFO host

2.7.2 *Use DMA for Data Transfer*

DMA uses FPGA memory to store data and then transfer it at high speed to host processor memory with very little processor involvement. This uses much fewer processor cycles when sending large blocks of data compared to the front panel indicator with FPGA Read/Write method (see Fig. 2.83).

Here are some programming instructions for implementing DMA:

- When setting the FPGA buffer size, you can use the default size (1023). Creating a larger FPGA memory buffer typically does **not** have benefits.
- You should set the host buffer size to a large value than the default size. By default, the host buffer size is 2 times bigger than the FPGA buffer. You should actually set it to at least two times the **Number of Elements** you plan to use.
- If you are passing an array of data, the **Number of Elements** input should always be an integer multiple of the array size. For example, if you are passing an array of 8 elements, the **Number of Elements** should be an integer multiple of 8 (such as 80, which would give 10 samples of 8 elements each.)
- Each DMA transaction has overhead, so reading larger blocks of data is typically better. The **DMA FIFO.Read function** automatically waits until the **Number of Elements** you requested become available, minimizing processor usage.
- Packing 16-bit channel data into a U32 (since DMA uses U32 data type) typically does **not** have benefits on CompactRIO, because the PCI bus has very high bandwidth for sending DMA data, so you most likely are nowhere near to using up all of the bus bandwidth. Instead, it is typically the processor that is the bottleneck in processing the data being streamed. Packing the data in the FPGA means it has to be unpacked on the processor, adding additional processor overhead. In general, you should send each channel as a U32 even if you are acquiring 16-bit data.
- The **Full** output on the DMA FIFO Write function is actually an error indicator. Under normal operation this should never occur so it is recommended that you stop the application if this error occurs and reset the FPGA before restarting.

2.7.3 Use the Minimum Data Type Necessary

Remember to use the minimum data type necessary when programming in LabVIEW FPGA. For example, using an 32-bit integer (I32) to index a Case Structure is probably overkill since it is unlikely that you will be writing code for **2 billion different cases**. Usually, an unsigned 8-bit integer (U8) does the trick, since it works for up to 256 different cases (see Fig. 2.84).

2.7.4 Optimizing for Size

The FPGA application shown in Fig. 2.85 is too large to compile, because it uses an array to store sine data.

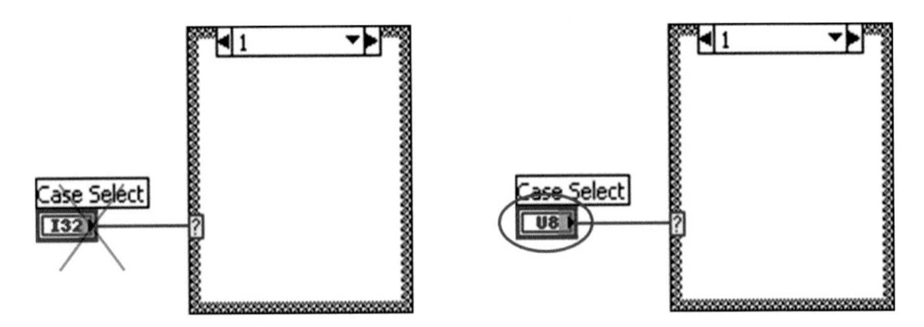

Fig. 2.84 Unsigned 8-bit integer

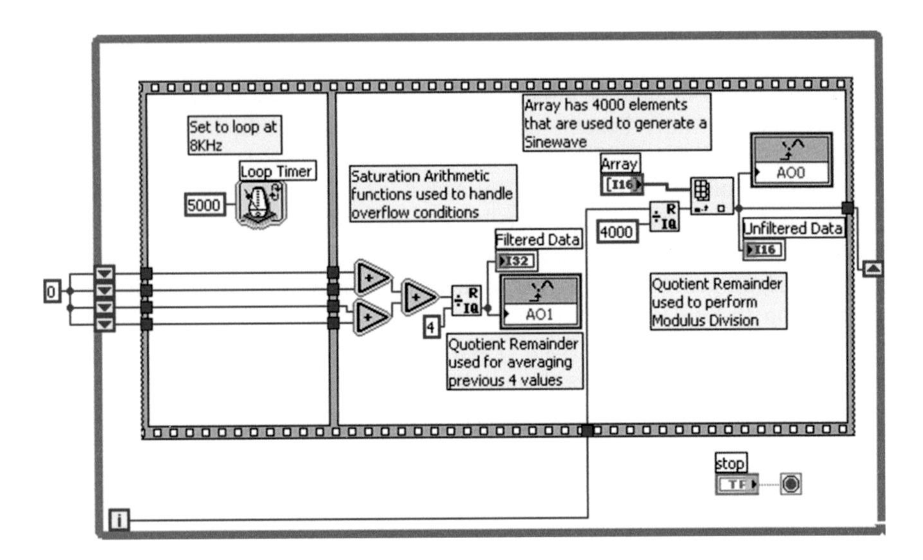

Fig. 2.85 Array to store sine data

The array is indexed to get a value. In addition, four previous points are stored in shift registers. The previous four values are averaged. This VI is too large to compile. What can be done to help optimize this code?

Gate hogs found in the code: Large front panel arrays, Remainder, and Quotient functions.

To improve the application, we replace the array with a look-up table as shown below (see Fig. 2.86).

This change alone has allowed us to compile the program and now it uses only 18 % of a 1 million gate FPGA. Can we further optimize the program?

Next we remove both Quotient Remainder (QR) functions. One of the QR functions was being used to index through the look-up table. This was replaced by a shift register counter operation. This shift register counter operation is a very

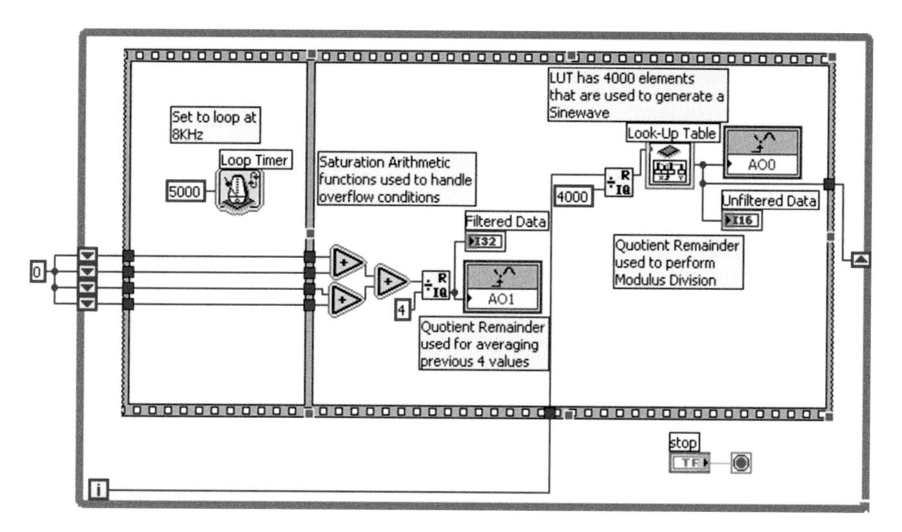

Fig. 2.86 Look-up table to store sine data

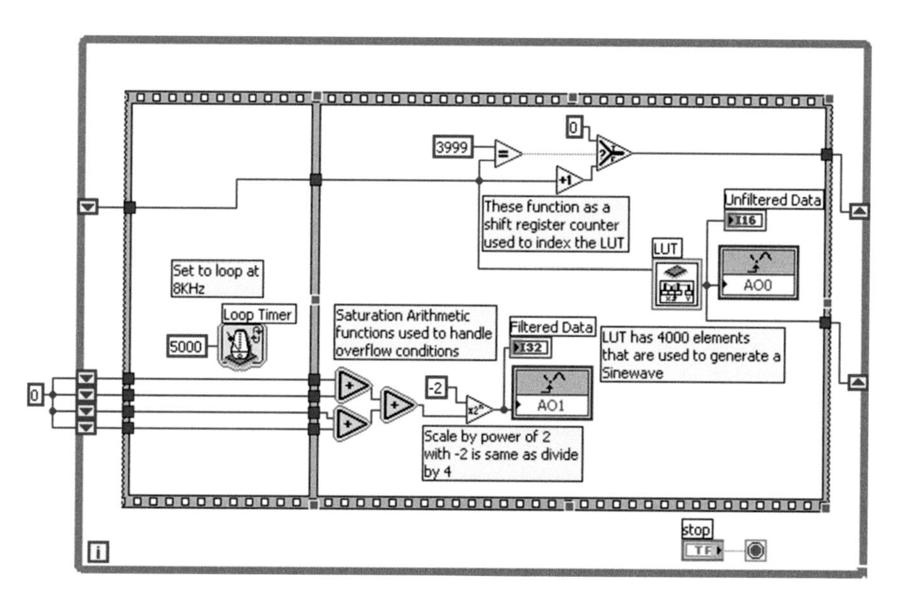

Fig. 2.87 Removed both quotient remainder functions

common technique in FPGA. The other QR function was replaced by a scale by 2 to the power of n. Because the scale by 2 has a constant input, it uses very little FPGA space (see Fig. 2.87). Note: Scale by 2^−2 is equal to dividing by 4.

Now the application takes only 9 % of the FPGA gates.

Table 2.4 Optimization methods for LabVIEW FPGA

Optimization technique	FPGA speed	FPGA size
Reduce *combinatorial paths*	✓	
Use *pipelining* when appropriate	✓	
Use *single-cycle timed loops*	✓	✓
Use *parallel operations*	✓	
Select appropriate *arbitration options*	✓	✓
Use *non-reentrant subVIs*		✓
Use *reentrant subVIs*	✓	
Limit the number of *front panel objects*, such as arrays		✓
Use the *smallest data type* possible	✓	✓
Avoid *large VIs and functions*, if possible	✓	✓
Schedule timing *using handshaking signals*	✓	✓

2.7.5 Additional Techniques to Optimize Your FPGA Applications

For more information on this topic, see the "Optimizing FPGA VIs for Speed and Size" topic on the NI Developer Zone [10]. The document contains detailed information on more than ten techniques you can use to optimize your LabVIEW FPGA applications. Table 2.4 shows some optimization methods for LabVIEW FPGA that can be implemented in your code.

References

1. http://www.ni.com/
2. http://sine.ni.com/nips/cds/view/p/lang/en/nid/210566
3. http://www.ni.com/example/7781/en/
4. http://www.ni.com/ipnet/
5. http://www.ni.com/fpga-hardware/whatsnew/
6. http://sine.ni.com/nips/cds/view/p/lang/en/nid/11766
7. http://sine.ni.com/nips/cds/view/p/lang/en/nid/11834
8. http://www.ni.com/compactrio
9. http://zone.ni.com/devzone/cda/tut/p/id/7781
10. http://zone.ni.com/reference/en-XX/help/371599D-01/lvfpgaconcepts/optimizing_fpga_vis/

Chapter 3
Real-Time Fuzzy Logic Controllers

3.1 Basic Parts in Real-Time Fuzzy Logic Controllers

Nonlinear control involves a nonlinear relationship between the controller's inputs and outputs and is more complicated than linear control; however, it is able to achieve better performance than linear control for many real-world control applications. Nonlinear control theory requires more challenging mathematical analysis and design than does linear control theory. An FLC is a nonlinear controller, that is, the function $f(\mathbf{x})$ is nonlinear. What distinguishes an FLC, T1 or T2, from other nonlinear controllers is that it generates its nonlinear mapping function $f(\mathbf{x})$ through linguistic if-then rules and linguistic terms for the antecedents and consequents of the rules (e.g., Low Speed, High Speed). Such rules can be (easily) obtained from a human operator or can be postulated and learned from data. According to Kosko [1], an FLC is unique in that it ties vague words like Low and High, and common sense rules, to state-space geometry [2]. Fuzzy logic controllers have two important advantages over other classes of nonlinear controllers.

- They are able to incorporate linguistic terms in the designs of the input–output membership functions.
- They are capable of handling uncertainties in inputs and state measurements more effectively.

Several fuzzy control systems nearly always run on non-real-time systems. If they do not require fast speed, they can run on conventional non-real-time systems. They may be able to run simple control systems with few difficulties. In addition, there are some applications that can run on non-real-time systems. Examples of non-control online real-time applications include supervision systems, detection systems, and fault detection. But now that FPGAs are faster than other devices, they can be implemented in different control systems in real time. Another advantage of real-time FPGAs is parallel execution. Most fuzzy FPGA systems run in parallel mode, much faster than sequential programs do. Since parallel programs do not

© Springer International Publishing Switzerland 2016

P. Ponce-Cruz et al., *Fuzzy Logic Type 1 and Type 2 Based on LabVIEW™ FPGA*, Studies in Fuzziness and Soft Computing,
DOI 10.1007/978-3-319-26656-5_3

backtrack, it is not necessary to keep all data in memory; this means that parallel systems can run indefinitely without overflowing memory. So real-time FPGA are great candidates for implementing fuzzy controllers. Typically, a block diagram of an industrial fuzzy control in real-time system can be represented by four main parts that are presented below [3]. Those parts have to be programmed in the FPGA.

- A rule base that is a set of if-then linguistic rules. In this part the fuzzy logic quantification of the expert's linguistic description about the controller is included.
- An inference mechanism that emulates the expert's decision-making in interpreting and applying knowledge about how best to control the plant.
- A fuzzification interface, which converts controller inputs into information that the inference mechanism can easily use to activate and apply rules.
- A defuzzification interface, which converts the conclusions of the inference mechanism into actual inputs for the process.

It is well known that in control programs utilizing the standard type of fuzzy control rules, which the number of rules goes up exponentially with the number of input variables, the "combinatorial explosion" problem.

This is because if we have *N* input variables fuzzified into *N* linguistic variables (discrete fuzzy sets describing numbers) with *M* linguistic terms (fuzzy set members) each, we construct rules with all combinations of the linguistic terms for each variable. This gives us M rules for one variable, M2 rules for two variables, and MN rules for N rules. If the objective of our program is to detect particular events in which we are interested, we may be able to use rules to detect the events in which we are interested and to ignore others, with a substantial reduction in the number of rules required. On the other hand, uncertainties that sometimes are non-avoided in control systems are:

- Uncertainties about a rule's consequent, when rules are obtained from a group of experts, because, as we have mentioned above, experts do not generally all agree on the same consequent.
- Linguistic uncertainties about the meanings of the words used in a rule's antecedent and consequent linguistic terms, because words mean different things to different people [4].
- Uncertainties associated with noisy training data that may be used to optimize (learn, tune) the MF parameters of an FLC.

3.2 Case Study: The Karnik–Mendel Algorithms Performance Implemented in Real-Time LABVIEW FPGA

Different research works present three KM (Karnik–Mendel) algorithms for the IT2FLS (interval Type 2 Fuzzy Logic Systems): The non-iterative [2] and the iterative [3] types that provide the same numerical results; the main difference is the improved total iteration count of the iterative type compared with the non-iterative one. Actually, both types are outperformed by the enhanced KM algorithm [5] which needs several initial conditions and lets the system converge faster than the other ones. These defuzzification methods get the generalized centroid of an IT2FLS. Different applications require opportune decisions; those decisions which require the noise immunity that the IT2FLS can provide [6]. The KM algorithms are very intensive and sometimes they are not appropriated for real-time applications. When they are dealing with a big amount of data, a late response appears generating an incorrect decision which cannot be acceptable in real-time hardware systems. The hardware implementation of IT2FLS offers good results but the software implementation is not good enough as it was shown in [7]. A hardware implementation is considered in this work for proving the KM algorithms in real-time applications.

For this purpose, a DC servomotor is analyzed. The IT2FLS is implemented in hardware by FPGA (field programmable gate array) based on LabVIEW and each method was implemented and tested independently. Two hardware considerations were taken: the first one is the swiftness and the second one is the final resource utilization. Every algorithm was compared using these two hardware conditions; also this paper shows a complete chart that presents the complexity of every algorithm according to the discrete discourse universe points.

This paper defines a new procedure of establishing the hardware implementation based on LabVIEW FPGA for KM algorithm. According to the results presented, various KM algorithms are not appropriated for real-time control applications. Finally, the expert must take into account the ratio between speed and area.

3.2.1 Interval Type-2 Fuzzy Logic Systems

The IT2FLS topology is the same as the T1FLS (type-1 fuzzy logic systems). It provides the fuzzification, the inference and the defuzzification stages.

The IT2FLS fuzzification maps the crisp values x_j into several membership values according to its membership degree. This means that a crisp value could belong to more than one IT2FS (interval type-2 fuzzy sets); those two membership degrees form a FOU (footprint of uncertainty). Each membership obtained after de fuzzification process is related using fuzzy logic operations as conjunction (AND) and disjunction (OR). The fuzzy conjunction allows the expert relating all the implied premises (input sets) and the fuzzy disjunction allows the expert aggregating all these implied values in order to obtain a specific consequent (output set). With all these relations a rule set is built. The rule set represents the IT2FLS conventional configuration.

Figure 3.1 shows a complete picture of the IT2FLS which can be used in hardware implementations. Each block can be an independent hardware entity. The inferred set is calculated and stored in a memory location.

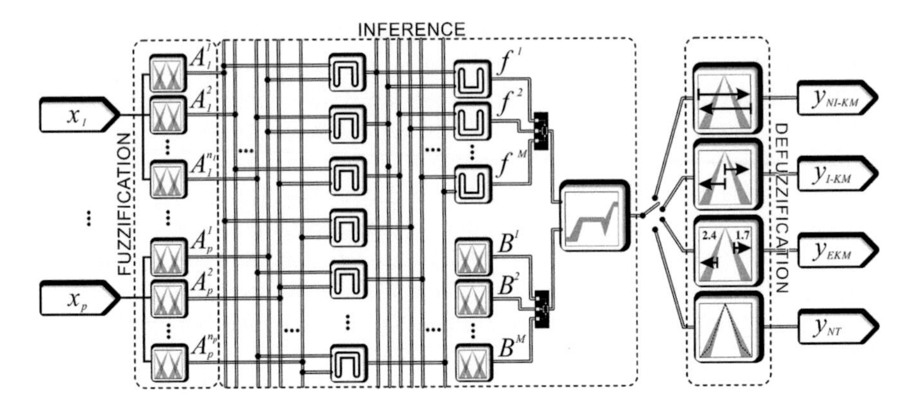

Fig. 3.1 Interval type 2 fuzzy logic system with four defuzzification methods: the non-iterative, iterative and enhanced KM algorithms, and the Nie-Tan method

3.2.2 The Karnik–Mendel Algorithm

In previous works [2, 3, 5, 8], the KM algorithm includes several modifications in order to decrement the impact of the large number of iterations required to generate a single centroid. This method is computationally exhaustive, different research works proposed some initial conditions and modifications based on the seminal algorithm. These modifications have the purpose of decreasing the search space for both the left and the right centroids.

Assume that the output discourse universe Y is a set of all the possible crisp outputs that can be obtained from a defuzzification method, i.e., $y \in Y$. Now let B be a consequent set defined along Y. The key of finding each approximated centroid is to find a switch point where the ES (embedded set) will change from one outer membership function to another; specifically, the Lower Membership Function (LMF) or $\underline{\mu}_{\tilde{B}}(y)$ the Upper Membership Function (UMF) or $\overline{\mu}_{\tilde{B}}(y)$, which comprises the FOU. Figure 3.2 provides a graphical description of the KM algorithm in general.

3.2.3 Non-iterative Version

The non-iterative KM algorithm [2] creates all the embedded sets θ_L and θ_R as possible from Eq. 3.1 to N. For finding the left and right centroids, the expert must calculate the following equations:

$$\theta \begin{cases} \overline{\mu}_{\tilde{B}}(y_i) & y_i \leq L \\ \underline{\mu}_{\tilde{B}}(y_i) & \text{otherwise} \end{cases}$$

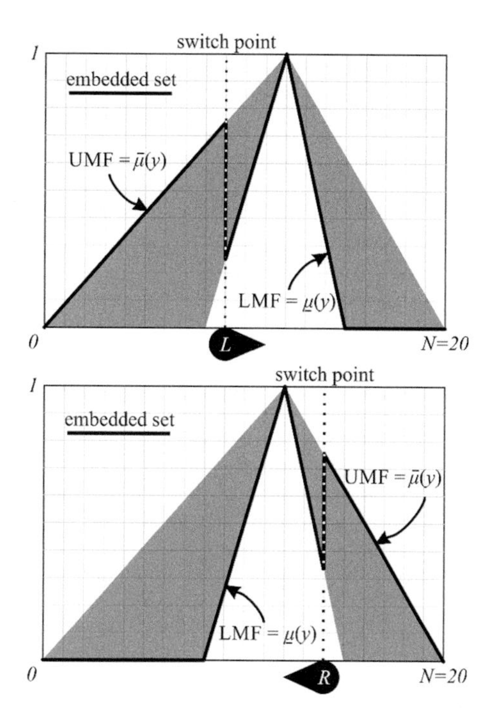

Fig. 3.2 The general centroid search in the KM algorithms. There is no centroid search for this algorithm; all the centroids are computed. This method is known as the non-iterative KM algorithm

$$c_{\theta_L} = \frac{\sum_{i=1}^{N} y_i * \mu_{\theta_L}(y_i)}{\sum_{i=1}^{N} \mu_{\theta_L}(y_i)} \tag{3.1}$$

where, $c_{\theta_L} \in \Theta_l$

$$\begin{aligned} c_l &= \min(\Theta_l) \\ c_{\theta_L} &= \frac{\sum_{i=1}^{N} y_i * \mu_{\theta_L}(y_i)}{\sum_{i=1}^{N} \mu_{\theta_L}(y_i)} \\ \theta_R &= \begin{cases} \underline{\mu}_{\tilde{B}}(y_i) & y_i \leq R \\ \bar{\mu}_{\tilde{B}}(y_i) & \text{otherwise} \end{cases} \\ c_{\theta_R} &= \frac{\sum_{i=1}^{N} y_i * \mu_{\theta_R}(y_i)}{\sum_{i=1}^{N} \mu_{\theta_R}(y_i)} \end{aligned} \tag{3.2}$$

where,$c_{\theta_R} \in \Theta$

$$c_r = \max(\Theta_r)$$

This non-iterative version creates a vector of left centroids Θ_l and a vector of right centroids Θ_r; from the left centroid vector, the far left centroid (the minimum) is selected while from the right centroid (maximum) vector, the far right centroid is

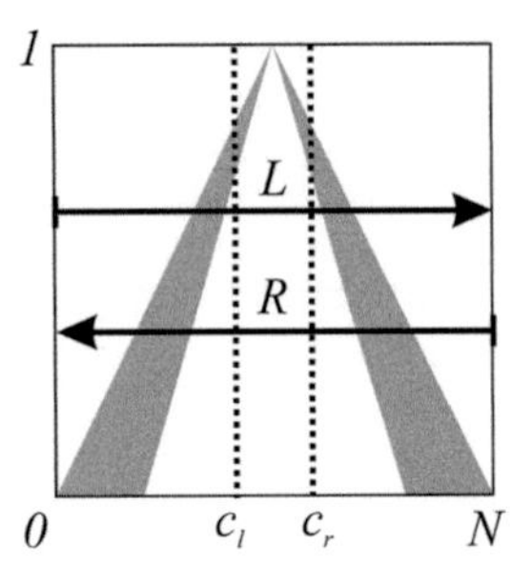

Fig. 3.3 The non-iterative KM algorithm performs all the centroid calculation

selected. Figure 3.3 shows the non-iterative version. Equations 3.1–3.2 can be rewritten as two summations in the numerator and the denominator as follows:

$$c_{\theta_L} = \frac{\sum_{i=1}^{L} y_i * \bar{\mu}_{\tilde{B}}(y_i) + \sum_{i=L+1}^{N} y_i * \underline{\mu}_{\tilde{B}}(y_i)}{\sum_{i=1}^{L} \bar{\mu}_{\tilde{B}}(y_i) + \sum_{i=L+1}^{N} \underline{\mu}_{\tilde{B}}(y_i)} \tag{3.3}$$

where, $c_{\theta_L} \in \Theta_l$

$$c_l = \min(\Theta_l)$$

$$c_{\theta_R} = \frac{\sum_{i=1}^{R} y_i * \underline{\mu}_{\tilde{B}}(y_i) + \sum_{i=R+1}^{N} y_i * \bar{\mu}_{\tilde{B}}(y_i)}{\sum_{i=1}^{R} \underline{\mu}_{\tilde{B}}(y_i) + \sum_{i=R+1}^{N} \bar{\mu}_{\tilde{B}}(y_i)} \tag{3.4}$$

where, $c_{\theta_R} \in \Theta_r$

$$c_r = \max(\Theta_r)$$

3.2.4 Iterative Version

This iterative version [3, 9, 10] searches for the left and right centroids starting from a convenient initial embedded set θ_i. After the initial centroid, the following centroid search of θ_i for the left and right centroids helps the algorithm to converge to the final centroid faster than the non-iterative version. Figure 3.4 shows the iterative KM algorithm. For the KM algorithm, the procedure to find the left and right centroid c_l and c_r is the following:

- (step 1) Sort all the discourse universe values y_i in ascending order, where $i = 1, 2, \ldots, N$, such $y_1 \leq y_2 \leq \ldots \leq y_N$. Associate each y_i with its corresponding $\underline{\mu}_{\tilde{B}}(y_i)$ and $\bar{\mu}_{\tilde{B}}(y_i)$.
- (step 2) Initialize

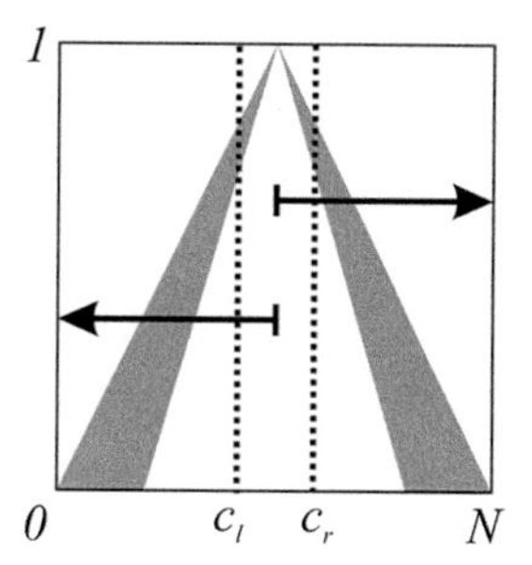

Fig. 3.4 The iterative KM algorithm. This algorithm is optimized for starting the centroid search from almost the half of discourse universe

$$\theta_i = \frac{\underline{\mu}_{\tilde{B}}(y_i) + \overline{\mu}_{\tilde{B}}(y_i)}{2}$$

and compute

$$y = \frac{\sum_{i=1}^{N} y_i \theta_i}{\sum_{i=1}^{N} \theta_i} \tag{3.5}$$

- (step 3) Find the switch point k, such $1 \leq k \leq N - 1$ and $y_k \leq y_{k+1}$,
- (step 4) Establish

For the left centroid:

$$= \begin{cases} \overline{\mu}_{\tilde{B}}(y_i) & i \leq k \\ \underline{\mu}_{\tilde{B}}(y_i) & \text{otherwise} \end{cases}$$

For the right centroid:

$$= \begin{cases} \underline{\mu}_{\tilde{B}}(y_i) & i \leq k \\ \overline{\mu}_{\tilde{B}}(y_i) & \text{otherwise} \end{cases}$$

and compute

$$y = \frac{\sum_{i=1}^{N} y_i \theta_i}{\sum_{i=1}^{N} \theta_i}$$

- (step 5) If $y' = y$. If true, then stop and assign $c_l = y$ or $c_r = y$, correspondingly, else continue.
- (step 6) Assign $y = y'$, y go to step 3.

In both versions, the generalized centroid is calculated simply by averaging both c_i and c_r centroids:

$$y = \frac{c_l + c_r}{2} \tag{3.6}$$

3.2.5 Enhanced Karnik–Mendel Algorithm

To improve the KM algorithm calculations, the enhanced KM algorithm [5] does not start only at the initial embedded set; it also starts from a convenient switch point (See Fig. 3.5).

The algorithm to find the left centroid c_i is the following:

- (step 1) Sort all the discourse universe values x_i in ascending order, where $i = 1, 2, \ldots, N$, such $y_1 \leq y_2 \leq \ldots \leq y_N$. Associate each y_i with its corresponding $\underline{\mu}_{\tilde{B}}(y_i)$ and $\overline{\mu}_{\tilde{B}}(y_i)$.
- (step 2) Establish $k = \text{round}(N/2.4)$ and compute

$$a = \sum_{i=1}^{k} y_i \overline{\mu}_{\tilde{B}}(y_i) + \sum_{i=k+1}^{N} y_i \underline{\mu}_{\tilde{B}}(y_i)$$

$$b = \sum_{i=1}^{k} \overline{\mu}_{\tilde{B}}(y_i) + \sum_{i=k+1}^{N} \underline{\mu}_{\tilde{B}}(y_i)$$

$$y = \frac{a}{b}$$

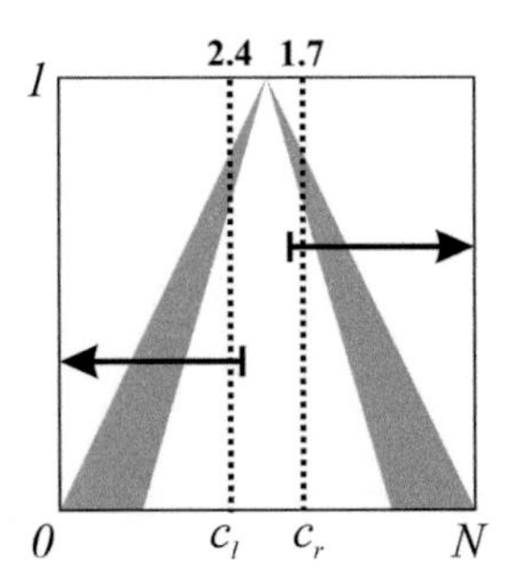

Fig. 3.5 The enhanced KM algorithm. This algorithm is doubly optimized for starting the centroid search at 8 % around N/2 of discourse universe. That is why some literature use the values 1.7 and 2.4 to start the centroid search

- (step 3) Find the switch point $k' \in [1, N-1]$, such

$$y_{k'} \leq y \leq y_{k'+1} \tag{3.7}$$

- (step 4) Checke if $k' = k$. If true, stop and assign $c_l = y$; else continue.
- (step 5) Compute

$$a' = a \pm \sum_{i=\min(k,k')}^{\max(k,k')} y\left[\overline{\mu}_{\tilde{B}}(y_i) - \underline{\mu}_{\tilde{B}}(y_i)\right]$$

$$b' = b \pm \sum_{i=\min(k,k')+1}^{\max(k,k')} \left[\overline{\mu}_{\tilde{B}}(y_i) - \underline{\mu}_{\tilde{B}}(y_i)\right]$$

and compute again

$$y' = \frac{a'}{b'}$$

- (step 6) Assign $y = y'$, $a = a'$, $b = b'$, $k = k'$ and go to step 3.

Each preliminary definition of $k_l = N/2.4$ and $k_r = N/1.7$ suggests that the initial centroids can be found ±8 % around the middle of the set support, i.e., $k_l = (N/2) - 0.08N$ and $k_r = (N/2) + 0.08N$. These values were obtained experimentally [5]. This initial search reduces the search of the final centroid.

3.2.6 Nie-Tan Method

Although the NT (Nie-Tan method) is not a KM algorithm derivation, this is used in the iterative version KM algorithms to find the initial switching point. This method is discussed in this work because of its simplicity. The simplicity of the algorithm is based on the average between the LMF and UMF as it is shown in Fig. 3.6.

The NT [11–13] is the simplest and fastest defuzzification method for IT2FLS. It searches the middle MF θ_i between UMF and LMF. Its centroid is the approximated generalized centroid:

$$\theta_i = \frac{\overline{\mu}_{\tilde{B}}(y_i) + \underline{\mu}_{\tilde{B}}(y_i)}{2} \tag{3.8}$$

$$y = \frac{\sum_{i=1}^{N} y_i * \mu_\theta(y_i)}{\sum_{i=1}^{N} \mu_\theta(y_i)} \tag{3.9}$$

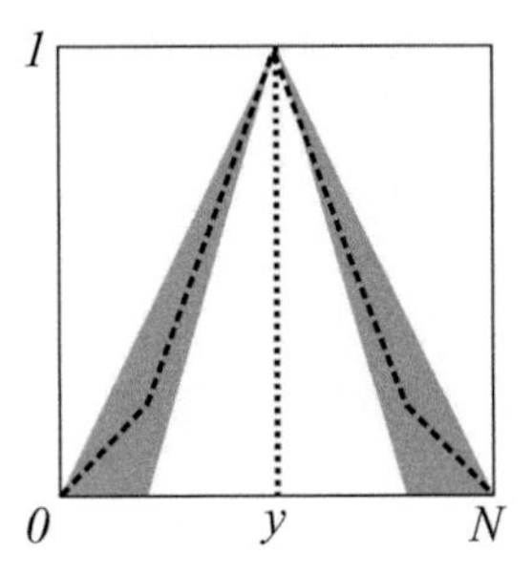

Fig. 3.6 The Nie-Tan method

This method is one of the fastest methods because it requires only a division by 2 (1-position right shift register) and only a single centroid calculation.

Following section explains a hardware overview in FPGA in which the KM algorithms and the NT were implemented.

3.3 DC Servomotor

This section presents a hardware study between several defuzzification methods for IT2FLS. For doing this, the DC Servomotor application is proposed. The following section provides some theory about the DC servomotor plant which is controlled by the IT2FLS. The DC servomotor model is well defined by conventional differential equations according to Fig. 3.7. This model contains the complete description of the servomotor. Normally noise in the control of DC servomotor appears from the sensors, so a controller that could rid of the noise effect is able to improve the whole performance of the system. The following figure shows the basic topology of the DC motor model.

The equations that describe the DC motor performance are presented below.

$$T = K_e * i \tag{3.10}$$

$$e_a = K_f * w_m \tag{3.11}$$

$$T = J * \frac{\mathrm{d}}{\mathrm{dt}} w_m + b * w_m \tag{3.12}$$

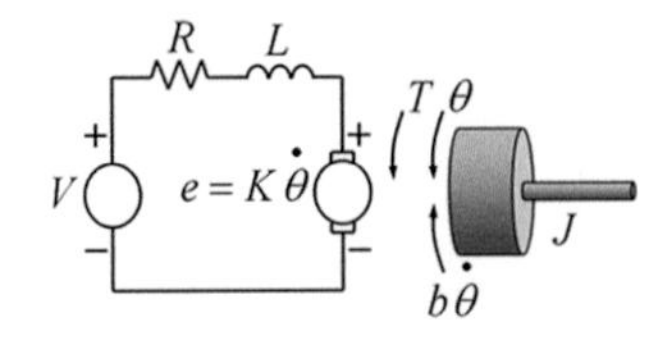

Fig. 3.7 DC servomotor model

$$L * \frac{\mathrm{d}}{\mathrm{dt}} i + R * i = V - e_a \tag{3.13}$$

where,

T: motor torque;
K_e: torque constant;
K_f: voltage constant;
i: armature current;
e_a: generated voltage;
w_m: angular velocity;
J: inertia;
b: damping ratio;
L: armature inductance;
R: armature resistance;
V: input voltage.

3.3.1 Laplace Transform Model

Rearranging Eqs. 3.10–3.13:

$$K * i(t) - b * w_m(t) = J * \frac{\mathrm{d}}{\mathrm{dt}} w_m(t) \tag{3.14}$$

$$V(t) - K * w_m(t) - R * i(t) = L * \frac{\mathrm{d}}{\mathrm{dt}} i(t) \tag{3.15}$$

Our principal goal is to control the DC servomotor speed by changing the input voltage V, so the Eqs. 3.14–3.15 can be transformed in terms of voltage V and angular speed w_m.

$$i(t) = \frac{J}{K} * \frac{\mathrm{d}}{\mathrm{dt}} w_m(t) + \frac{b}{K} * w_m(t) \tag{3.16}$$

Replacing TT 3.16 in Eq. 3.15 we get,

$$\left(\frac{L * J}{K}\right) * \frac{\mathrm{d}^2}{\mathrm{dt}^2} w_m(t) + \left(\frac{L * b + R * J}{K}\right) * \frac{\mathrm{d}}{\mathrm{dt}} w_m(t) + \left(\frac{K^2 + R * b}{K}\right) * W_m(t) = V(t) \tag{3.17}$$

Before applying Laplace transform it is necessary to represent the last equation in terms of deviation variables,

$$\left(\frac{L*J}{K}\right)*\frac{\mathrm{d}^2}{\mathrm{dt}^2}w_m(t)+\left(\frac{L*b+R*J}{K}\right)*\frac{\mathrm{d}}{\mathrm{dt}}w_m(t)+\left(\frac{K^2+R*b}{K}\right)*W_m(t)=\forall(t) \tag{3.18}$$

where,

$$W_m(t)=w_m(t)-w_m(0)$$
$$\forall(t)=V(t)-V(0)$$

and $w_m(0)$ and $V(0)$ are the initial conditions.

Applying the Laplace transform we get,

$$\left(\frac{L*J}{K}\right)*s^2w_m(s)+\left(\frac{L*b+R*J}{K}\right)*sW_m(s)+\left(\frac{K^2+R*b}{K}\right)*W_m(s)=\forall(s) \tag{3.19}$$

Rearranging Eq. 3.19, we obtain the transfer function that represents the servomotor plant in terms of Laplace transform as follows:

$$\frac{W_m(s)}{\forall(s)}=\frac{1}{\left(\frac{L*J}{K}\right)*s^2\left(\frac{L*b+R*J}{K}\right)*s+\left(\frac{K^2+R*b}{K}\right)} \tag{3.20}$$

3.3.2 State-Space Transfer Function

From Eq. 3.16, it can be defined the state-space model introducing the next variables:

$$x_1=w_m(t) \tag{3.21}$$

$$x_2=\frac{\mathrm{d}}{\mathrm{dt}}w_m(t) \tag{3.22}$$

The model is defined as,

$$x_1=\frac{\mathrm{d}}{\mathrm{dt}}w_m(t)=x_2 \tag{3.23}$$

$$x_2=\frac{\mathrm{d}}{\mathrm{dt}}w_m(t)=\left(\frac{K}{L*J}\right)*V(t)-\left(\frac{L*b+R*J}{L*J}\right)*x_2-\left(\frac{K^2+R*b}{L*J}\right)*x_1 \tag{3.24}$$

$$y = w_m(t) = x_1 \tag{3.25}$$

$$u = V(t) \tag{3.26}$$

Finally, the state-space equations are,

$$\begin{bmatrix} x_1 \\ x_2 \end{bmatrix} = \begin{bmatrix} 0 & 1 \\ -\left(\frac{K^2 + R*b}{L*J}\right) & -\left(\frac{L*b + R*J}{L*J}\right) \end{bmatrix} \begin{bmatrix} x_1 \\ x_2 \end{bmatrix} + \begin{bmatrix} 0 \\ \left(\frac{K}{L*J}\right) \end{bmatrix} [u] \tag{3.27}$$

$$y = \begin{bmatrix} 1 & 0 \end{bmatrix} \begin{bmatrix} x_1 \\ x_2 \end{bmatrix} \tag{3.28}$$

And the DC motor model constants are,

$$A = \begin{bmatrix} 0 & 1 \\ -\left(\frac{K^2 + R*b}{L*J}\right) & -\left(\frac{L*b + R*J}{L*J}\right) \end{bmatrix} \tag{3.29}$$

$$A = \begin{bmatrix} 0 \\ \left(\frac{N}{L*J}\right) \end{bmatrix} \tag{3.30}$$

$$A = \begin{bmatrix} 1 & 0 \end{bmatrix} \tag{3.31}$$

$$D = 0 \tag{3.32}$$

3.3.3 Servomotor Control System

The servomotor control systems for the defuzzification method performance comparison is shown in Fig. 3.8, where the IT2FLS performs the position control using the position error θ and the change of the position θ; also, the servomotor plant can be represented by Eq. 3.20 or Eqs. 3.27–3.28. Figure 3.9 presents some details about the IT2FLS and its rules; nine rules are performed relating the antecedents to infer the consequences.

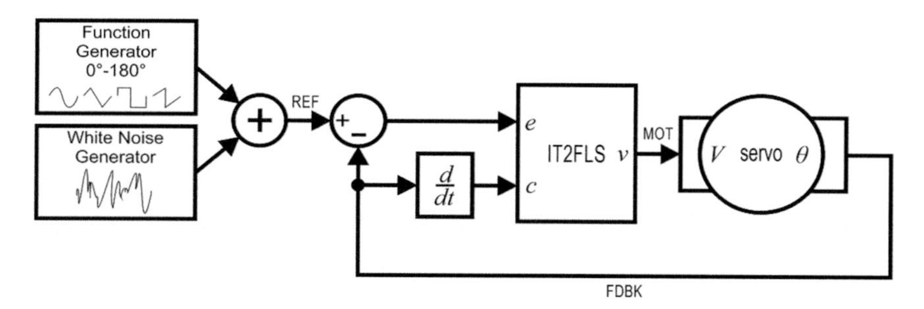

Fig. 3.8 The servomotor control system

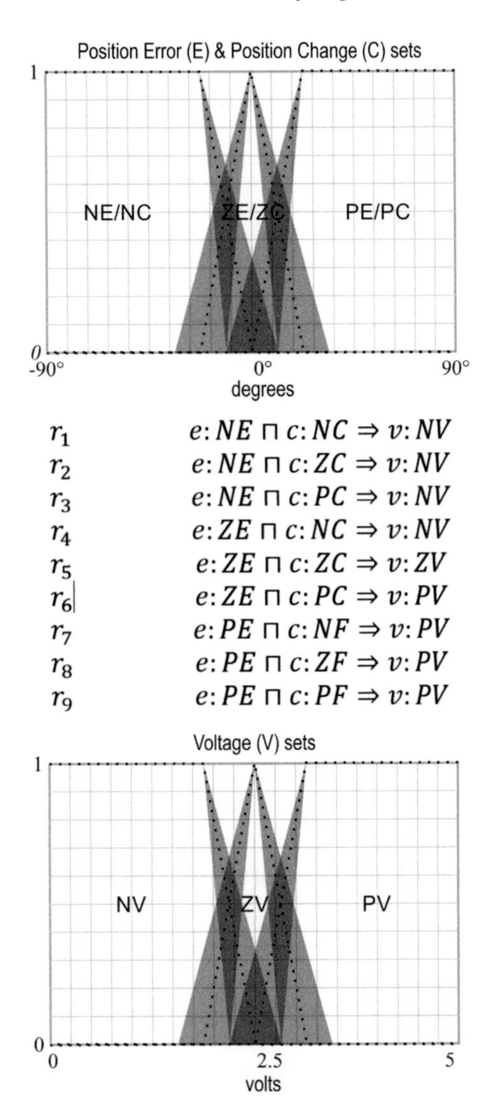

Fig. 3.9 The fuzzy sets and the rule set for the servomotor application

3.4 The Hardware Complexity

A computational device can manage a real-time application but the device selection will depend on the latency of that digital system that will manage it. A real-time application requires fast digital systems (with low latency) because its real-time characteristic will be defined by its WCET (worst-case execution time).

The centroid calculation had been considered as a problem in T1FLS, for real-time applications because of its large latency. Although several applications were solved without problems with the centroid calculation in T1FLS, the use of this in the KM algorithms are very common and intensive and therefore the IT2FLS are hugely more complex than the T1FLS. The search of each centroid for each embedded set is globally several times more complex compared with the T1FLS.

This is one of the reasons why the IT2FLS are not used as the T1FLS today, and its robustness capability is not approached. The IT2FLS are often implemented in hardware such as RT-MCU (real-time microcontroller units) or DSP (digital signal processors) [14] and FPGA [15], because the rest of the hardware (computers with sequential program execution) becomes impractical and not feasible for real-time applications. These digital systems are often more expensive, especially the FPGA, and commonly this economic reason limits the application implementation in most cases.

Some of the high-end FPGA hardware elements that can be found in real-time applications are:

- high-speed dedicated multiplications;
- high-speed dedicated memory;
- DSP blocks;
- real parallelism;
- partial reconfiguration.

Full-customizable multiplications are available in the FPGAs as a solution for improving the timing performance. Xilinx is the main FPGA brand that is used widely in several applications along the world. Some Xilinx FPGA families provide faster multiplication blocks and additional high-performance resources (like DSP blocks and RT processors in the same land) in their higher end devices. Some high-speed RAM (Random Access Memory) blocks are available for storing data, for instance.

These resources help the expert to design high-speed and low-cost T1FLS and IT2FLS in hardware among the software approaches, because a sequential program is executed in a fixed computing architecture [7]. If the specific application deserves the use of IT2FLS, then the expert must take into account how to select the appropriated defuzzification method that will be used for its real-time application.

3.5 Methodology

The hardware defuzzification method comparisons are performed implementing all the KM algorithms, the NT method in FPGA hardware; National Instruments Reconfigurable I/O (compact RIO or cRIO) device is used and LabVIEW FPGA module to program every defuzzification stage is enunciated in this paper and all the IT2FLS architecture. The cRIO device used is the NI-9014 with analog I/O capabilities, the C-series NI-9263 and the NI-9201 modules.

For this reason, the expert may also know some hardware details about the KM algorithms that will use. For hardware comparison, the DC servomotor application is proposed as a study case where the rotor position θ is tracked when a signal generator is introduced and a noise generator is used to disturb the IT2FLS performance, according to Fig. 3.8. Although the servo case is relatively slow, all these defuzzification methods can be analyzed and compared. Several studies have been implemented in FPGA hardware for T1FLS [16] and IT2FLS [17]. However, no comparisons have been presented, in hardware terms, where the most important defuzzification methods are applied for solving the same problem.

The DC servomotor application, as can be seen in Fig. 3.8. It is an electronic control training module dedicated for the cRIO device where the user can test control performance easily. The servomotor moves the rotor position from 0° to 180° according to the REF analog signal; the REF signal is used in the cRIO as an input to determine the reference that the controller will follow. The feedback (FDBCK) signal is another analog signal used in the cRIO as an input also; the difference is that this signal is used to know where the current rotor position is located.

The control system described in Fig. 3.8 is implemented using the LabVIEW FPGA module [14], where several diagram blocks are used for the data acquisition (input nodes), the noise generator summed to the reference signal (signal generator), the data driving to the servo (output node), and finally the IT2FLS. The error is computed as the difference between the desired values, such the reference signal (REF), i.e., the reference and the current position value FDBCK as can be seen in Fig. 3.8.

The cRIO analog output is used to drive the desired voltage to the servomotor that is calculated using the fuzzy logic controller presented in Fig. 3.9. This fuzzy logic controller determines the new position of the rotor. This signal is applied in the motor input voltage terminals signal (MOT).

Details about the servomotor control training module are not known, although their general equations are defined (Eqs. 3.20, 3.27–3.28). This is not important for the fuzzy controller, because the IT2FLS must determine its control law without knowing the exact system dynamics.

For comparison purposes, the noisy reference signal REF is used so that the IT2FLS can follow it, the fuzzification and inference processes are previously tuned and are the same for the four cases. Four signal shapes were tested with the four methods: the Non-Iterative Karnik–Mendel Algorithm (NI-KM), the Iterative Karnik–Mendel Algorithm (I-KM), the Enhanced Karnik–Mendel Algorithm (EKM), and the Nie-Tan Method (NT), as can be seen in Fig. 3.10.

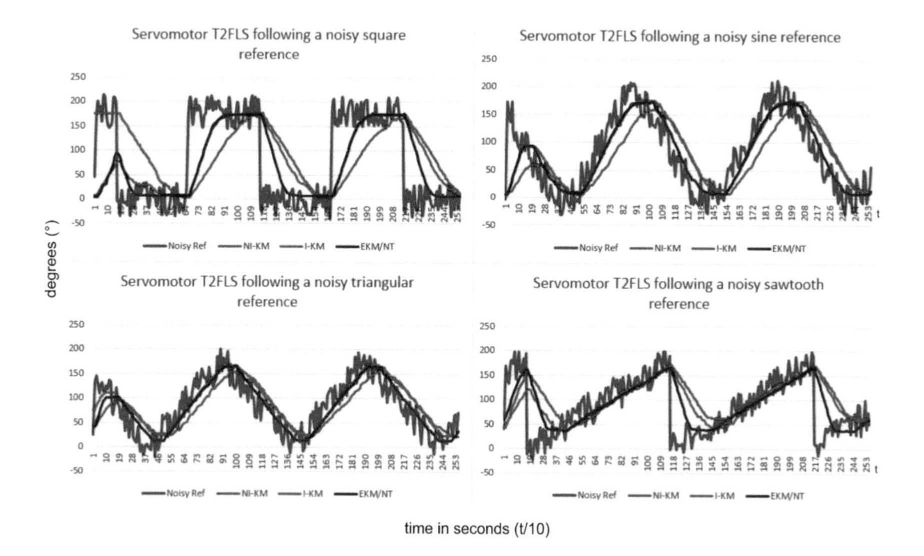

Fig. 3.10 All the KM algorithms and NT method tracking performance comparison in the presence of noise. Four signal shapes were used to be tracked by the IT2FLS using each algorithm. The total time for each experimental test is 25.5 s

3.6 Results and Discussion

3.6.1 Reference Tracking

The final response of the control system using all the defuzzification methods mentioned is shown in Fig. 3.10. Four signal shapes were used for the IT2FLS tracking.

The non-iterative KM algorithm is the worst case and its latency makes the decision to take late decisions. The best results were obtained with the EKM algorithm and the NT method. This behavior is similar in each case.

3.6.2 The Hardware Performance

The following paragraphs are dedicated to describe the hardware complexity between every defuzzification method described in the previous sections. NI cRIO devices are used in conjunction with NI LabVIEW FPGA module.

The following section provides a timing and area performance analysis where each KM algorithm is analyzed and compared. Also, the NT method is included.

3.6.2.1 Complexity and Arithmetic Related

Assume that each algorithm requires two memories with N locations and each centroid (either left or right) are computed in parallel. Each element is a byte (8-bit width). Every memory can be distributed in the FPGA by LUT (Look-Up Tables).

The following chart provides a comparison between the entire KM algorithms and the NT method.

According to Eqs. 3.1–3.8, the total iteration count can be obtained, as an approximation if a counter is included in the inner loop of the algorithm and by observation of the block diagram. In this case, $N = 256$, where N is the number of points the discourse universe is divided into. So, the non-iterative KM may last $2N^2 + 4N = 2(256)^2 + 4(256) = 132.096$ iterations for a single defuzzification without taking into account the iterations required for the T2 fuzzification and T2 inference processes, as can be seen in Table 3.1. For instance, the EKM requires from 1 up to $0.1764N^2 + 2.1N = 0.1764(256)^2 + 2.1(256) = 12.098$ iterations to find the final left or right centroid. Due to each left or right centroid calculation processes in parallel; it can be seen as a single loop. The Iterative version, which can be compared with the Non-iterative version and may last the same time to execute with the difference of the early termination condition, converges slower than the EKM.

The NT method also finishes before, because its implementation requires only a single centroid calculation and only a single embedded set in the set FOU. Although the NT method seems to be the fastest, the EKM may find the final centroid in the first iteration, which is very significant compared to the NT algorithm.

Table 3.1 summarizes the complexity of each defuzzification method, which was obtained experimentally.

3.6.2.2 Resource Usage

The cRIO device provides several limited resources like multiplications, memories, amongst others, which provide very high performance, and the expert may use to build more complex structures like divisions or square roots. LabVIEW FPGA

Table 3.1 Number of iterations per defuzzification method

Element/method	Non-iterative KM	Iterative KM	EKM	NT
ES calculation	N	N	N/A	N
ES centroid calculation	$N + 3$	$N + 3$	$N + 1$	$N + 3$
Total iteration count	$2N^2 + 4N$	$[N/2, 2N^2 + 4N]$	$[1–0.1764N^2 + 2.1N]$	$2N + 3$

module is a GUI (Graphical User Interface) which provides several basic tools that can be used to implement high-throughput operations in a RIO device, controls that let the user modify the input data to the digital system and indicators that let the user show the output data from the digital system, related to a virtual instrument that works as the programming unit.

In functional block terms (LabVIEW FPGA module), the KM algorithms and the NT method were designed, implemented are compared according to the structures used for being implemented.

After the module compilation, LabVIEW FPGA provides the FPGA resulting resource utilization. As can be seen in the last chart (Table 3.2), the resource usage can be useful for comparison purposes. This way, it can be seen that the EKM is the most expensive, but this is the fastest defuzzification method, but the NT method is the cheapest method because it requires only one division and one multiplication for calculating the final centroid. Because this method does not search for several embedded sets, the final centroid is a single value and that is why it requires a single centroid calculation unit. The EKM and the I-KM require three centroid calculation units, while the non-iterative version requires two.

3.6.2.3 Timing and Area Resource Usage

Also, the final timing and resources of each method is presented using the LabVIEW Tick Count block. And the resource utilization is obtained from the Build Specifications in LabVIEW FPGA module. The best timing performance is reached by the EKM algorithm and the second best one is the NT, although the EKM also presents the worst resource utilization. The non-iterative KM algorithm presents the worst case, achieving about 20 Fuzzy Logic Inferences Per Second (FLIPS), which is not practical for the IT2FLS applied in real-time applications. Also, the NT method presents the best resource utilization. The software VI timing performance depends on the operating system tick time, which is generally 55 ms per tick, so for the NI-KM the total ticks are 1.537, then 1.537×55 ms = 84.535 ms, as can be seen in Table 3.3.

Table 3.2 Number of hardware elements used for each defuzzification method

Structure/method	Non-iterative KM	Iterative KM	EKM	NT
Multiplications	1	1	6	1
Divisions	1	2	5	1
Sums/subtracts	3	4	17	5
Centroid calculation units	2	3	3	1
Comparator/multiplexers (MUX)	5	9	9	1

Table 3.3 Timing performance and resource utilization per defuzzification method

Resource/methods	Non-iterative KM	Iterative KM	EKM	NT
Latency (hardware) in milliseconds	49.48	0.8875	0.1756	0.27
Latency (software) in milliseconds	84.535	1.87	1.43	1.32
Slices	1461	2415	2593	915
Registers	1454	2087	2828	959
LUT	2185	3759	3965	1305

References

1. B. Kosko, Fuzzy systems as universal approximators. Comp. IEEE Trans. **43**(11), 1329–1333 (1994)
2. J.M. Mendel, Type-2 fuzzy sets and systems: an overview. IEEE Comput. Intell. Mag. **2**(1), 20–29 (2007)
3. N.N. Karnik, J.M. Mendel, Q. Liang, Type-2 fuzzy logic systems. IEEE Trans. Fuzzy Syst. **7**, 643–658 (1999)
4. J.M. Mendel, Uncertain rule-based fuzzy logic system: introduction and new directions (2001)
5. H. Wu, J.M. Mendel, Enhanced Karnik-Mendel algorithms. IEEE Trans. Fuzzy Syst. **17**, 923–934 (2009)
6. M. Biglarbegian, W. Melek, J. Mendel, Robustness of interval type-2 fuzzy logic systems, in *IEEE NAFIPS*, Canada, 2010
7. N. Manaresi, R. Rovatti, E. Franchi, R. Guerrieri, G. Baccarani, Automatic synthesis of analog fuzzy controllers: a hardware and software approach. IEEE Trans. Ind. Electron. **43**(1), 217–225 (1996)
8. J. Mendel, X. Liu, Some extensions of the Karnik-Mendel algorithms for computing an interval type-2 fuzzy set centroid, in *IEEE Symposium on Advances in Type-2 Fuzzy Logic Systems*, Paris, France, 11–15 Apr 2011
9. J. Mendel, F. Liu, Super-exponential convergence of the Karnik-Mendel algorithms used for type-reduction in interval type-2 fuzzy logic systems, in *IEEE International Conference on Fuzzy Systems*, Vancouver, BC, Canada, 16–21 July 2006
10. J. Mendel, R.I.B. John, Type-2 fuzzy sets made simple. IEEE Trans. Fuzzy Syst. **10**(2), 117–127 (2002)
11. J.M. Mendel, X. Liu, Simplified interval type-2 fuzzy logic systems. IEEE Trans. Fuzzy Syst. **21**(6), 1056–1069 (2013)
12. D. Wu, J.M. Mendel, On the continuity of type-1 and interval type-2 fuzzy logic systems. IEEE Trans. Fuzzy Syst. **19**(1), 179–192 (2011)
13. M. Nie, W.W. Tan, Towards an efficient type-reduction method for interval type-2 fuzzy logic systems, in *IEEE World Congress on Computational Intelligence,* Hong Kong, China, 1–6 June 2008
14. R. Lauwereins, M. Engels, J.A. Peperstraete, Parallel processing enables the real-time emulation of DSP ASICs, in *IEEE International Workshop on Rapid System Prototyping*, North Carolina, USA, 4–7 June 1990
15. Y. Chen, V. Dinavahi, Multi-FPGA digital hardware design for detailed large-scale real-time electromagnetic transient simulation of power systems. IET Gener. Transm. Distrib. **7**(5), 451–463 (2013)
16. M. Cirstea, J. Khor, M. McCormick, FPGA fuzzy logic controller for variable speed generators, in *IEEE International Conference on Control Applications*, Mexico, 5–7 Sept 2001
17. M. Melgarejo, C.A. Pena-Reyes, Implementing interval type-2 fuzzy processors. IEEE Comput. Intell. Mag. **2**(1), 63–71 (2007)

Chapter 4
Fuzzy Logic Type 1 and Type 2 LabVIEW FPGA Toolkit

This chapter presents the toolkit developed for LabVIEW FPGA, so that it is possible to implement Fuzzy Logic systems in LabVIEW FPGA in a fast way by the Toolkit. The main blocks can be used for implementing complex fuzzy logic control systems that can be adjusted to different applications according to the user needs. The theoretical part was covered in chapter one; thus this chapter deals with the implementation of fuzzy logic systems.

4.1 Type-1 Fuzzy Sets

The FPGA Fuzzy Logic Toolkit developed has functional blocks for mapping crisp values to membership degrees. This mapping is performed by membership functions, as it was described in Chap. 1. A membership function, or simply MF, is a convex mathematical function (continuous or noncontinuous) defined over a discourse universe $\mathbb{R}:[-\infty, \infty]$, i.e., a real set of crisp values which works as the function domain, that maps their values to its codomain defined in interval $\mathbb{R}:[0, 1]$. Due to the digital representation of the fuzzy set, membership values are defined in 8-bits, i.e., $\mathbb{N}:[0, 255]$. Last definition is valid for continuous or discrete systems. In the FPGA devices, due to its discrete nature, only discrete MFs are allowed. The shape of the MF characterizes the fuzzy set. Hence, only for the fuzzy logic toolkit purposes, the FS will be called MF indistinctly and vice versa.

The available MFs in this toolkit for Fuzzy logic type 1 (T1FLS) are the following:

- Trapezoidal
- Triangular
- S-Shape
- Z-Shape

A MF can be built according to some basic parameters, just as the according-to-shape parameters, normalization values, and the uncertainty parameters (type 2). The next sections describe those parameters.

© Springer International Publishing Switzerland 2016

P. Ponce-Cruz et al., *Fuzzy Logic Type 1 and Type 2 Based on LabVIEW™ FPGA*, Studies in Fuzziness and Soft Computing,
DOI 10.1007/978-3-319-26656-5_4

4.1.1 Membership Function Parameters

A MF, as described in the first chapter, needs some parameters in order to know where it is located over its discourse universe. For example, a triangular shaped MF has three characteristic points which are the parameters that defines where it is located, its width, and its center. Every MF, regardless its shape, receives an input cluster with several unsigned 8-bit fixed point controls, which defines the MF parameters according to its shape. A Trapezoidal MF could have four characteristic points (a cluster with four unsigned 8-bit fixed point values) which defines how long its left or right shoulder is and how long is the minor base, i.e., the MF region where there are more than one maximal, values in its membership degree.

Table 4.1 shows the characteristic points of each available MF and their meaning.

Table 4.1 Available MF for T1FLS fuzzification

MF shape	Symbol	Graphical representation	Description
Trapezoid	 TRAP	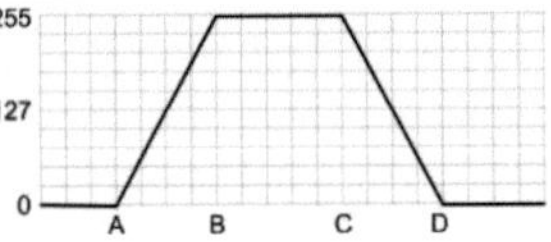 255 127 0 A B C D	Intervals [*A*, *B*] and [*C*, *D*] need slope calculation. In interval [*B*, *C*] several maximal membership values are present. The MF support is [*A*, *D*]; so for any other discourse value away from this interval the membership degree is equal to zero
Triangular	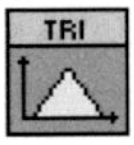 TRI	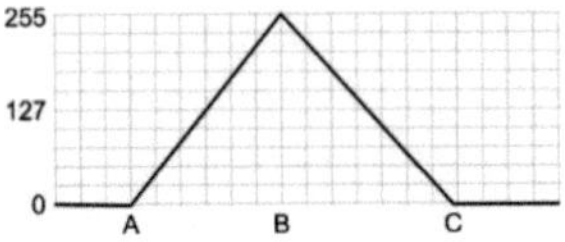 255 127 0 A B C	Intervals [*A*, *B*] and [*B*, *C*] need slope calculation. Unlike trapezoidal MF, there is a single maximal membership value in *B*. The MF support is [*A*, *C*]; so for any other discourse value away from this interval the membership degree is equal to zero
S-shape	 S	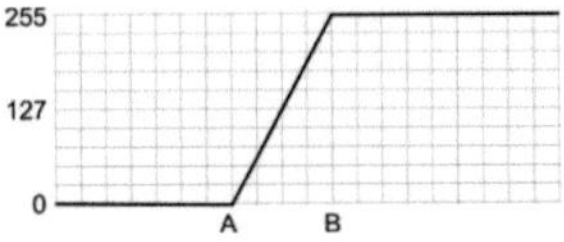 255 127 0 A B	Interval [*A*, *B*] need slope calculation. After *B*, for each discourse value there is a maximal membership value. Before *A*, every membership degree is equal to zero
Z-shape	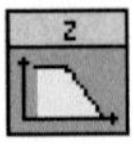 Z	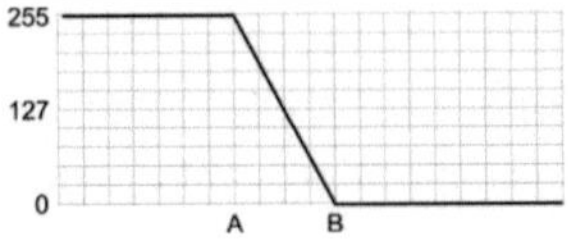 255 127 0 A B	Alike S-shape MF, interval [*A*, *B*] need slope calculation. Before *A*, for each discourse value there is a maximal membership value. After *B*, every membership degree is equal to zero

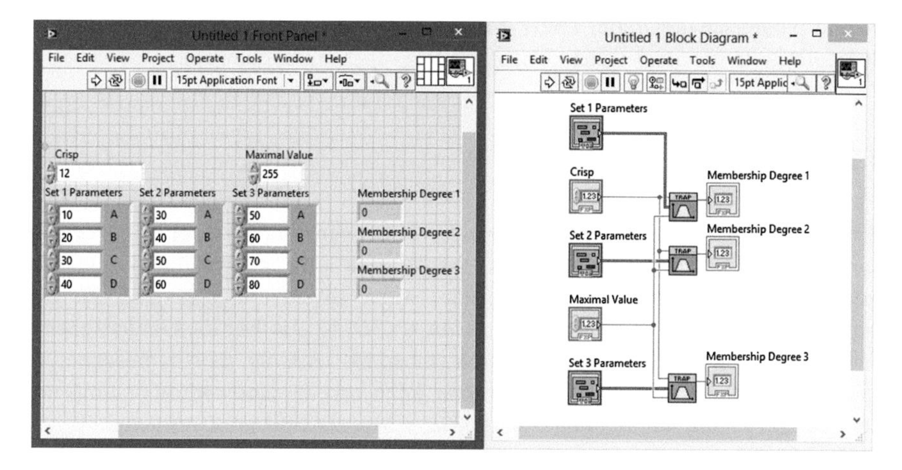

Fig. 4.1 Single variable fuzzification using trapezoidal MFs

Figure 4.1 shows a simple VI with several trapezoidal MFs. As you can see, the MF is four characteristic points as its parameters and its support is over interval $[A, D]$. Also, set overlapping is present, e.g., Set 1 and Set 2 are overlapped in intervals set 1: $[C, D]$ and set 2: $[A, D]$.

4.1.2 Normalization

When an MF whose maximal value is different from zero is called *unnormalized* MF, it is called *normalized if the maximum value is different from zero.*

Every MF has an unsigned 8-bit fixed point control that defines its maximal value; also each set may have a different normalization value.

4.1.3 Membership Degree

An unsigned 8-bit fixed point indicator is available for displaying the resulting membership degree.

4.1.4 Error Handling

If you specify an invalid parameter, i.e., if you violate the rule $A \leq B \leq C \leq D$, the MF will set a Boolean indicator. Some MFs may generate an infinite slope; e.g., for the triangular MF, if you make $A = B$, this can cause a very big slope, and then it is in error state. The MF has also an *Error* indicator, which may be useful to catch an invalid parameter definition. Every MF has an unsigned 8-bit fixed point control that defines its maximal value.

4.2 Type-2 Fuzzy Sets

Alike the T1FS, the FPGA fuzzy logic program has functional T2FS blocks. Unlike the T1FS, the T2FS has several MFs that describe a higher uncertainty level. In general, T2FS has a primary membership, where several T1 embedded sets are defined and forms a *Footprint Of Uncertainty* (FOU); also has a secondary membership, where a T1FS determines which embedded set is the most representative.

However, due to the reason that general T2FS and the hardware limitations are very complex, only the *Interval Type-2 Fuzzy Sets* (IT2FS) are considered.

The IT2FS maps the crisp input value from a variable and maps it to two membership degrees, usually the FOU boundary sets, i.e., the *Lower Membership Function* (LMF) and the *Upper Membership Function* (UMF).

Those two MFs are defined exactly as the T1FLS over a discourse universe $\mathbb{R}: [-\infty, \infty]$, and maps their values to two codomains, also defined in interval $\mathbb{R}: [0, 1]$. So, the primary membership is characterized by the boundary sets and the T1FS of the secondary membership is replaced by an interval set.

This strongly reduces the complexity of the system. From this moment, when we refer to a T2FS, we will refer it as IT2FS indistinctly.

Alike T1FLS, the available MF for T2FLS are the following:

- Trapezoidal
- Triangular
- S-Shape
- Z-Shape

These T2MFs are very similar to the T1MFs. The only difference is an additional cluster for the uncertainty width definition and the membership degrees are presented as a cluster of two indicators.

4.2.1 Membership Function Parameters

Alike the T1FS, the parameters are the same for each MF and they are available for tuning in a similar cluster of the T1FS.

4.2.2 Normalization

Unlike the T1FS, there is a cluster with two unsigned 8-bit fixed point numerical control that defines the maximal value of each LMF or UMF, i.e., their heights.

You must assure that the maximal value of LMF is lower or equal to the maximal value of UMF. Similarly, a unnormalized MF for a T2FS is defined when the maximal values of UMF or LMF are not equal to one.

4.2.3 *Uncertainty Widths*

For each characteristic point in the parameter cluster, there is an uncertainty value. This value lets the T2FS definition expanding the FOU around the characteristic point, i.e., a parameter value.

This helps to generate both LMF and UMF, where each one is defined according to the parameters and uncertainty values. If you specify these values to zero, then you will create a T1FS.

An additional cluster of four unsigned 8-bit fixed point controls are available for tuning. The only exception is the triangular MF, because this cluster has only two unsigned 8-bit fixed point controls, defined for parameters A and C; if you add an uncertainty value to parameter B, you will convert the triangular MF into trapezoidal MF.

Table 4.2 presents another four available MFs for the T2FS. Also, Fig. 4.2 shows a simple example for the T2FLS defuzzification, similar to the ones described for T1FLS.

4.2.4 *Membership Degrees*

The membership values obtained from the T2FS are available as a cluster of two unsigned 8-bit fixed point indicators, where the upper indicator is the value for the UMF and the lower for the LMF.

Table 4.2 Four conventional MFs for type 2 fuzzy system

Membership functions	(MFs)
Trapezoid	TRAP
Triangular	TRI
S-shape	S
Z-shape	Z

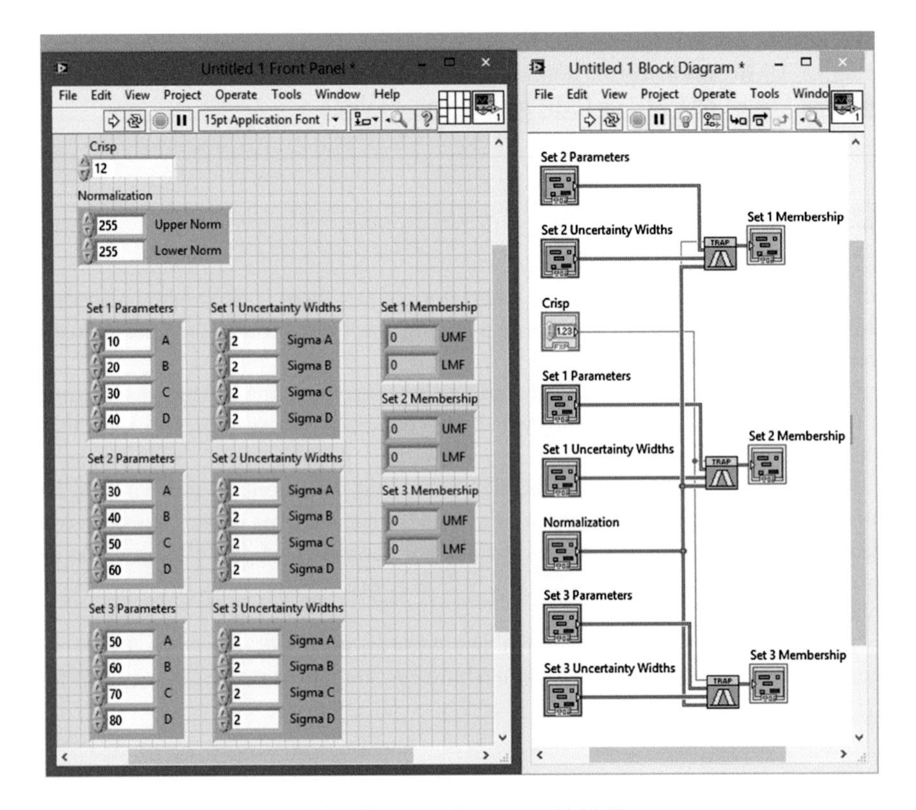

Fig. 4.2 Single variable type-2 fuzzification using trapezoidal MFs

4.2.5 *Error Handling*

According to the T1FS error handling, the T2FS error handling is exactly the same, because the T2FS are built with T1FS. So, if an invalid parameter set is caught then its Boolean indicator will set.

In both T1 and T2 FS, the fuzzification process starts when a crisp value, related to an input variable, is converted to a membership value. These membership values are used as premises to infer a consequent set.

In the toolkit, each input variable is related to a control (as you can see in Figs. 4.1 and 4.2) and every input value to the *Crisp* connection port is linked to it; so each set shares this control in order to overlap every MF. Also, in Figs. 4.3 and 4.4, each set is overlapped due to their parameter values. Table 4.3 depicts the membership functions included in the toolkit and a general description.

So, if you need to create two input variables, you will need two unsigned 8-bit fixed point numerical controls that feeds two groups of FS, each group for each input variable and every membership values obtained is related to their corresponding input variable (see Figs. 4.3 and 4.4).

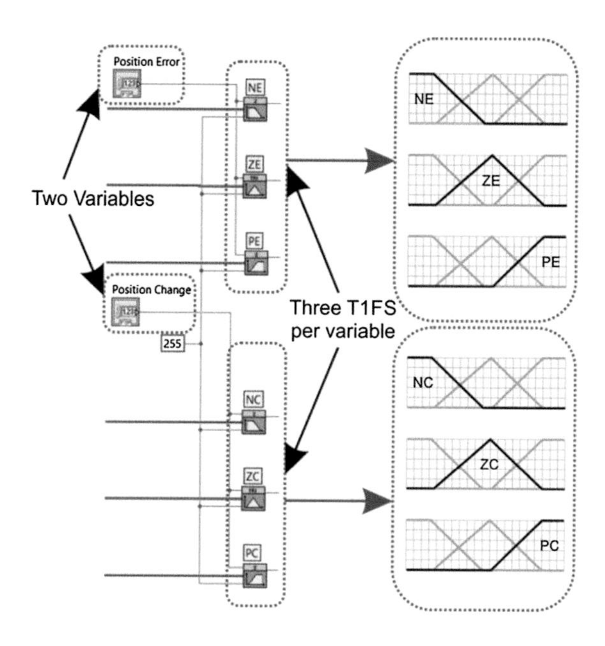

Fig. 4.3 A simple fuzzification stage with two input variables and three T1FS per variable characterized by T1MFs

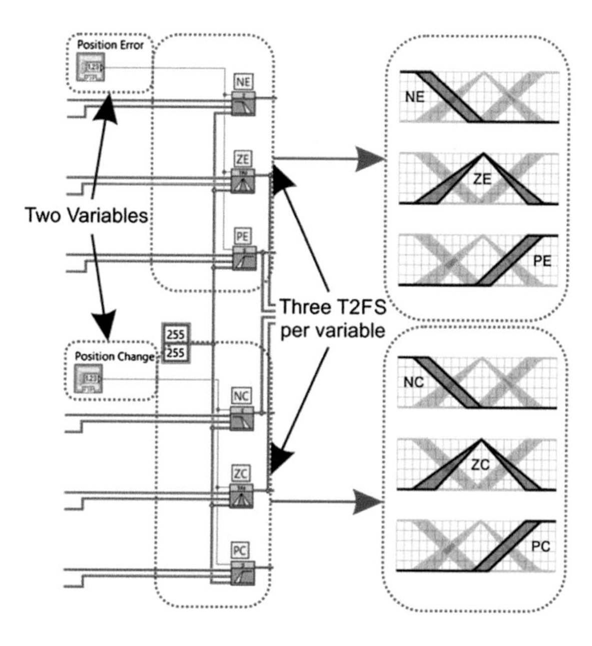

Fig. 4.4 A simple fuzzification stage with two input variables and three T2FS per variable characterized by FOUs

Table 4.3 Available MF for T2FLS fuzzification

MF shape	Symbol	Graphical representation	Description
Trapezoid	TRAP	255, 127, 0; ←B→←C→; ←A→ ←D→	There are two trapezoidal MFs that fulfill the same restrictions of the T1 trapezoidal MF, described in the last section
Triangular	TRI	255, 127, 0; ←A→ B ←C→	There are two triangular MFs that fulfill the same restrictions of the T1 trapezoidal MF, described in the last section. The value *B* has not an uncertainty value. Its uncertainty cluster is defined only for *A* and *C* parameters
S-shape	S	255, 127, 0; ←B→; ←A→	There are two S-Shape MFs that fulfill the same restrictions of the T1 trapezoidal MF, described in the last section
Z-shape	Z	255, 127, 0; ←A→; ←B→	There are two Z-Shape MFs that fulfill the same restrictions of the T1 trapezoidal MF, described in the last section

In the following program, the inference process is presented with some rule examples that show the easy way of building inference machines.

4.2.6 Examples

The FPGA fuzzy logic program includes two fuzzification Vis for each T1FLS and T2FLS located in the examples\chapter2 directory. These examples let you interact with all the parameters of the sets and see the response in the corresponding indicators.

1. Open a new project and add the following VI files:

   - T1 fuzzy set example.vi: This VI is useful for users who do not know how to create an input variable with several T1FS. The user can change the set parameters and verify from the indicators their corresponding membership values (see Fig. 4.5).
   - T2 fuzzy set example.vi: Similar to the last VI, you can create an input variable with several T2FS and change the set parameters and uncertainties, and verify from the indicators their corresponding membership values. In this case, you can verify that their indicators are clusters of two indicators (see Fig. 4.6).

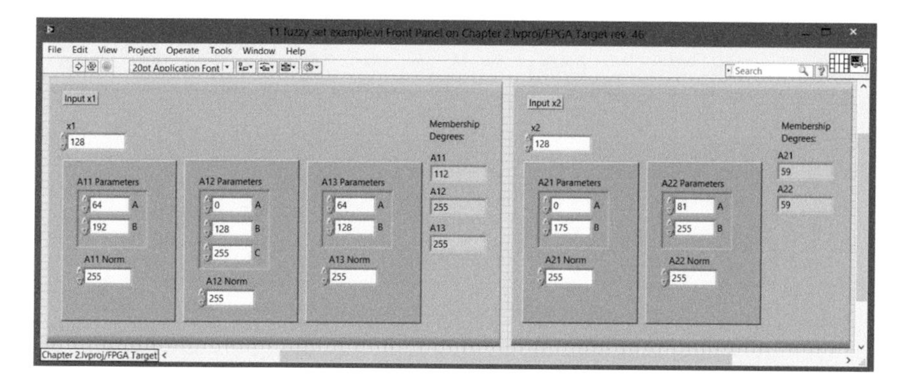

Fig. 4.5 The front panel of the T1 fuzzification stage: note that each variable has their corresponding FS and each FS has their corresponding membership degrees

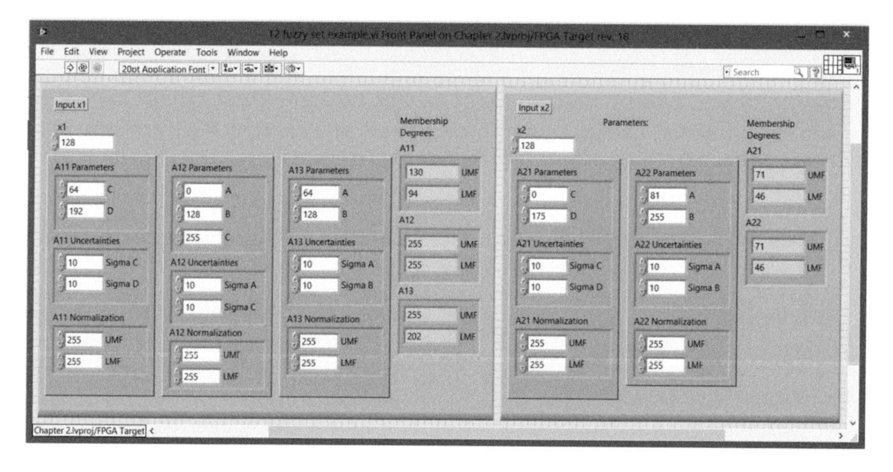

Fig. 4.6 The front panel of the T2 fuzzification stage: note that each variable has their corresponding T2FS and each T2FS has their corresponding membership degrees clustered. Each membership degree has two numerical indicators: UMF and LMF

- A single T1 fuzzy set example.vi: This VI provides a single T1 fuzzy set example, used for the Graph a single T1 fuzzy set example.vi.
- A single T2 fuzzy set example.vi: This VI provides a single T2 fuzzy set example, used for the Graph a single T2 fuzzy set example.vi.
- Graph a single T1 fuzzy set example.vi: This VI let the user interact with the Type-1 set parameters and normalization values and all the available shapes in the chapter. The user can verify graphically by means of a waveform chart its corresponding shape.
- Graph a single T2 fuzzy set example.vi: This VI let the user interact with the Type-2 set parameters, uncertainties and normalization values and all the

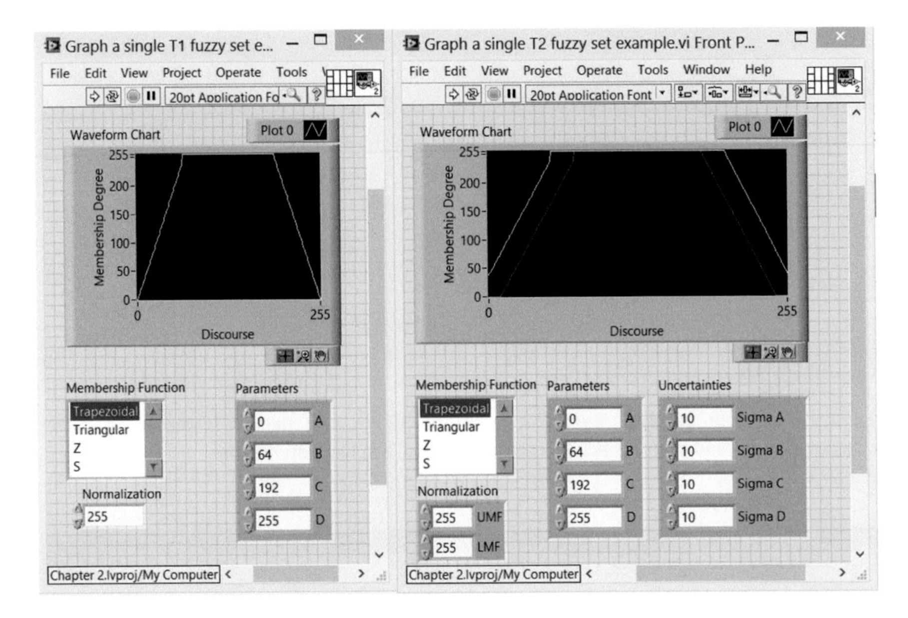

Fig. 4.7 This Vis let the user graph their T1 and T2 fuzzy sets according to their corresponding shape, parameters, uncertainty widths, and normalization values

available shapes in the program. The user can verify graphically by means of a waveform chart its corresponding shape (see Fig. 4.7).

2. Add three FIFOs in the FPGA target named MF, UMF, and LMF with the following characteristics:

   - Target-Host DMA FIFO,
   - Unsigned 8-bit fixed point data length, 8-bit for the integer part,
   - Do not change the memory access arbitration.

3. Compile the following Vis in the current target:

   - T1 fuzzy set example.vi
   - T2 fuzzy set example.vi
   - A single T1 fuzzy set example.vi
   - A single T2 fuzzy set example.vi

4. Now, you can execute and interact with the T1/T2 fuzzy sets and see how their shapes change whenever you change the parameter controls.

For additional information about how to detect the FPGA target and how to create a FPGA project, please refer to Chap. 2.

4.3 Creating a Knowledge Base

4.3.1 Building a Rule Set

A linguistic rule that uses the form IF-THEN can be built with the functional blocks available in the FPGA fuzzy logic toolkit. The set of rules is constructed as was presented in Chap. 1.

The following procedure can be followed in order to build a knowledge base:

- First, define the number of FS of each input variable,
- Define the number of rules. The number of possible rules is the multiplication of the total number of FS in each input variable. It does not mean that you have to define all the rules.
- For each variable, every FS must be connected to the same crisp input and every FS parameter must lie in the same discourse universe of that variable.
- Imply the premises. All the premises must be related with a conjunction operation. You have to select a FS from a variable and connect it to one input of the conjunction operation and select another FS from another variable and connect it to the other input. If three premises will be implied, then you have to add an additional conjunction operation where its first input must be connected to the output of the first conjunction operation; and the third premise must be connected to the second input of the second conjunction operation. This array is known as cascade (For type-1 rules see Fig. 4.8; for type-2 see Fig. 4.9).
- Aggregate the contributing premises. Every rule consequence must be aggregated with a disjunction operation, always that two or more rules contribute to the same consequence. If more than one rule contributes to the same rule, then you have to follow the same cascade array described in step 4 (Also, for type-1 rules see Fig. 4.8; for type-2 see Fig. 4.9).
- Calculate the *Inferred Set* (IS). The final value obtained from the aggregation is called the *Firing Strength*. This determines the influence strength that a rule or a group of rules have. Each firing strength is useful for cutting the output consequent sets and helps the system contribute in a specific decision related to that consequence. If you perform a minimum intersection between the consequent set in question and the firing strength, you will get its real FS shape or MF. After the firing strengths are used, you must add all the active consequent sets using a maximum union.

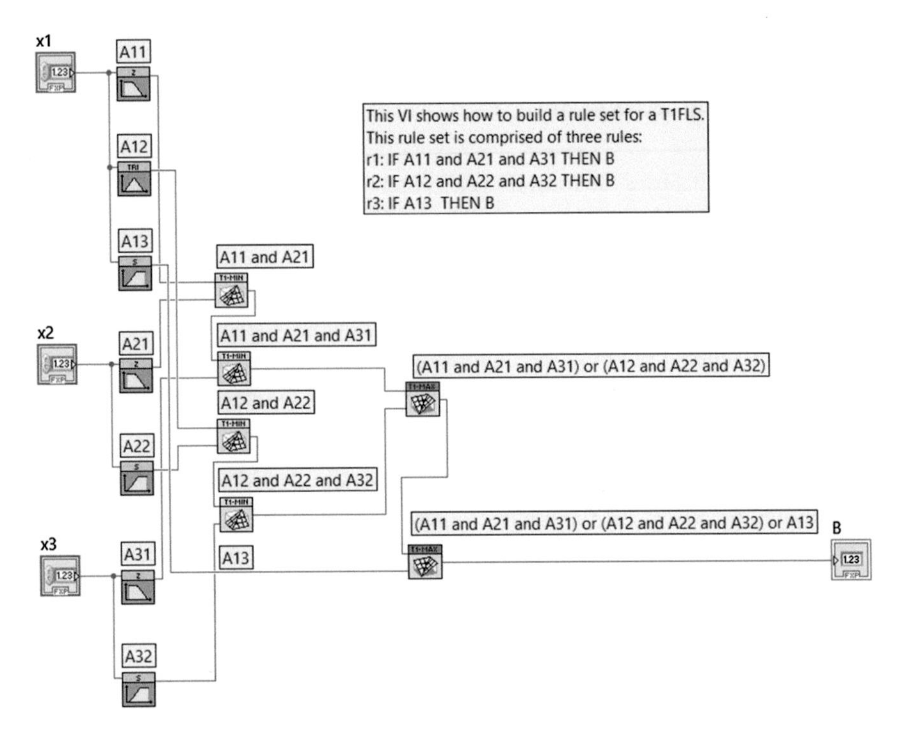

Fig. 4.8 T1 inference machine. Note that the fuzzification stage is included. Each control in the VI makes that all the FS generate a membership degree. That membership degree is related using the minimum intersection according to the rules 3.1–3.3. All the rules are aggregated using the maximum union. Indicator *B* are the corresponding T1 firing strength of consequent set *B*

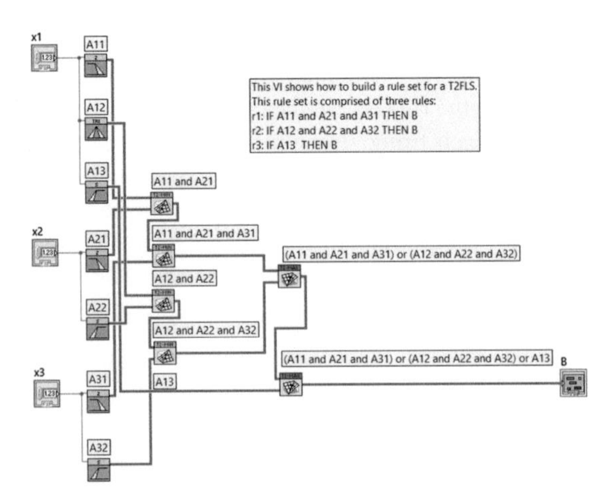

Fig. 4.9 T2 inference machine. Note that the fuzzification stage is included also. Each control in the VI makes that all the T2FS generate two membership degrees. UMF and LMF. Those membership degrees are related using the minimum intersection according to the rules 3.1–3.3. All the rules are aggregated using the maximum union. Cluster *B* are the corresponding T2 firing strength of consequent set *B*

4.4 The Inferred Set

The rules process FSs in order to get a specific decision. The output of the inference machine is also a FS, according to its type, i.e., if the inference machine processes T1FSs, its output must be a T1FS. Also, if the inference machine processes T2FSs, its output must be a T2FS, i.e., the *Inferred Set* (IS).

Sometimes, it is useful to obtain the IS to calculate the generalized centroid for searching the type-reduced set (for T2FLS) or defuzzification. Some defuzzification methods or type reducers require the IS shape (see Fig. 4.10).

The FPGA fuzzy logic toolkit includes two functional blocks that calculate the IS from a T1 or a T2 inference machine.

Consider the rules in Eqs. 3.1–3.3. In this example, there are three input variables with the following sets: $x_1 : A_1^1, A_1^2, A_1^3$, $x_2 : A_2^1, A_2^2$, and $x_3 : A_3^1, A_3^2$. They also have a single consequent set B.

Rule 1 implies three premise sets: A_1^1, A_2^1, and A_3^1; one per input variable. Each implication is performed with a conjunction operator. Let us select two minimum intersection operators for this purpose. Rule 2 implies three premises also, i.e., A_1^2, A_2^2, and A_3^2; and finally rule 3 implies only two premises, i.e., A_2^1 and A_3^2 (step 4).

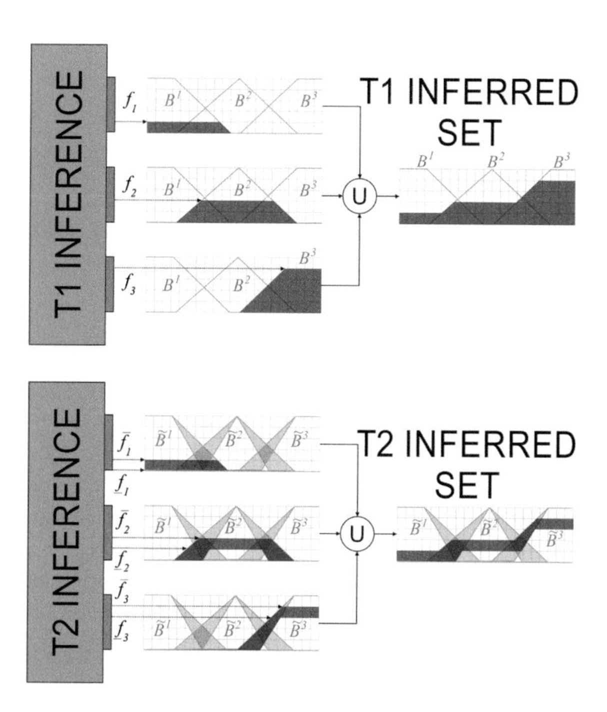

Fig. 4.10 How to connect the firing strengths to a T1 inferred set functional block

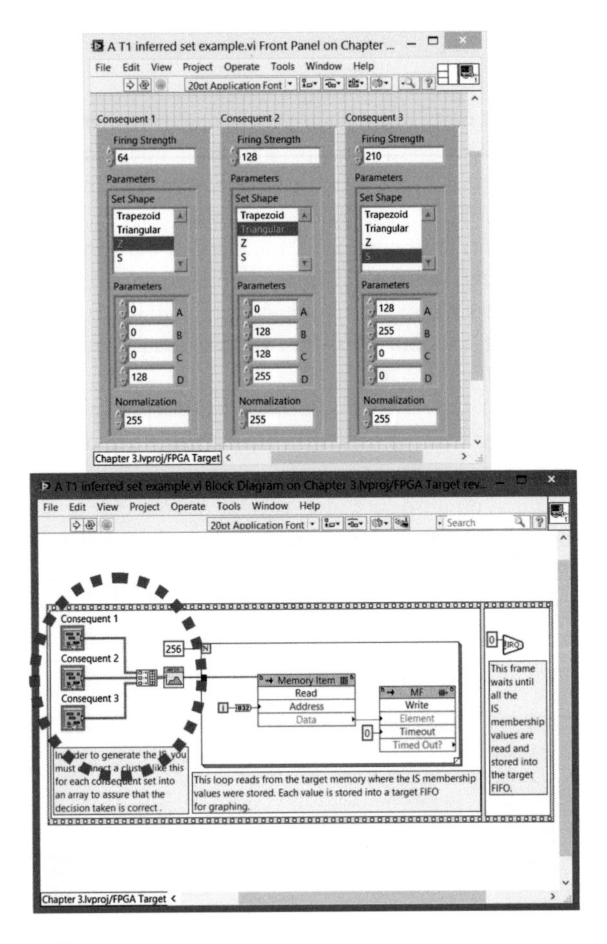

Fig. 4.10 (continued)

As the three rules have as consequence the set B, their results (contributions) are aggregated using a disjunction operator; let us select two maximum union operators for this purpose (step 5).

Those values obtained after the inference step are the firing strengths, as stated before. With the firing strengths and the consequent set MF, you can build the IS. Observe that in Figs. 4.13 and 4.14, each firing strength trims their corresponding consequent set; so the union of all those trimmed consequent sets is the *inferred set.*

Now, please refer to Figs. 4.10 and 4.11. There are two target VIs provided for user comprehension about the inferred set calculation: a T1 inferred set example.vi and a T2 inferred set example.vi. They are part of the execution of a host VI called Graph a T1 inferred set example.vi and Graph a T2 inferred set example.vi.

For both T1 and T2IS, some specifications must be stated. So, the T1 inferred set and T2 inferred set functional blocks,

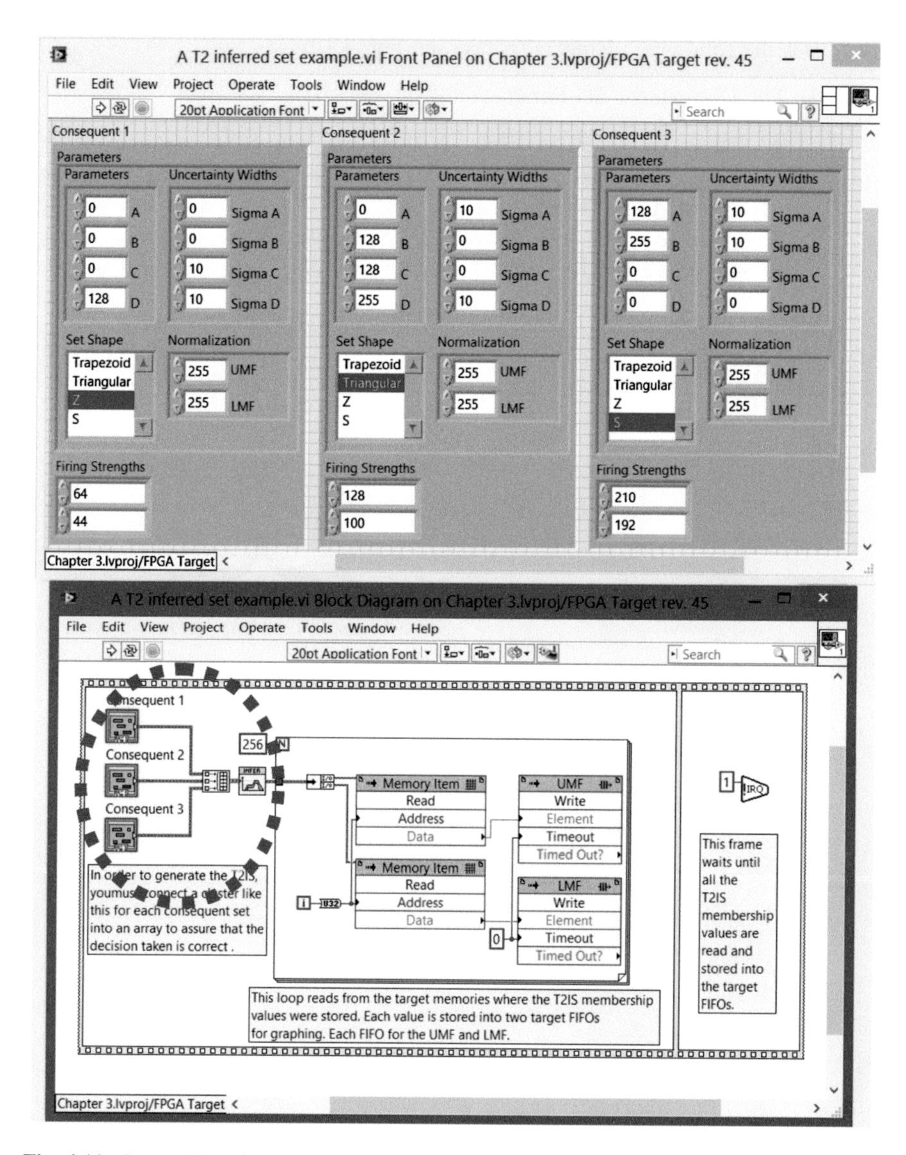

Fig. 4.11 Connection of the firing strengths to a T2 inferred set functional block

- Can calculate an inferred set for a single output. So, if you need two output variables you must add two of these blocks to your VI.
- Can process up to 10 consequent sets for a single output.
- Have a single input port (control), i.e., a 10-element fixed array of clusters, one element per consequent set. So, each array cluster element has the following characteristics:

– Has the following elements (please see Fig. 4.12):

1. The Consequent Set Parameters

   - Shape: As stated in Chap. 2, there are four MFs available in the chapter (Z-shape, Triangular, Trapezoidal, and S-shape). They are presented in a listbox.
   - Parameters: For both T1 and T2FLS, the set parameters correspond to each characteristic point of the selected shape in the listbox. Elements

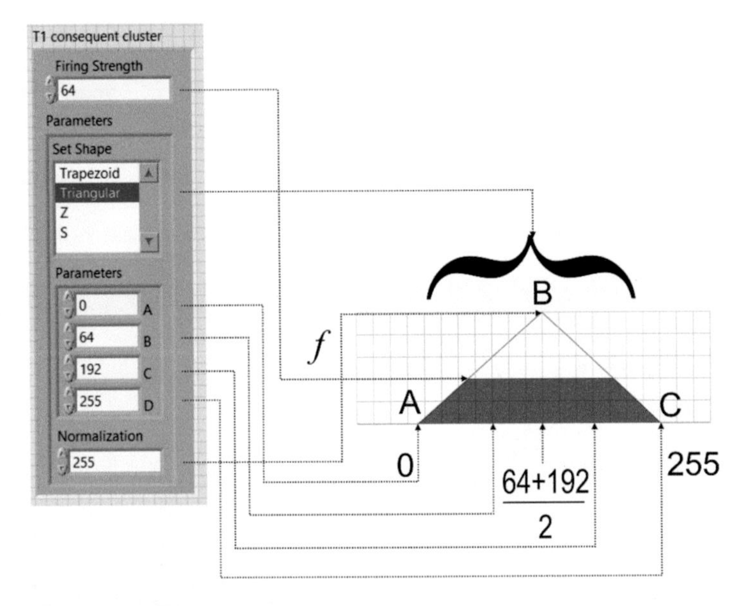

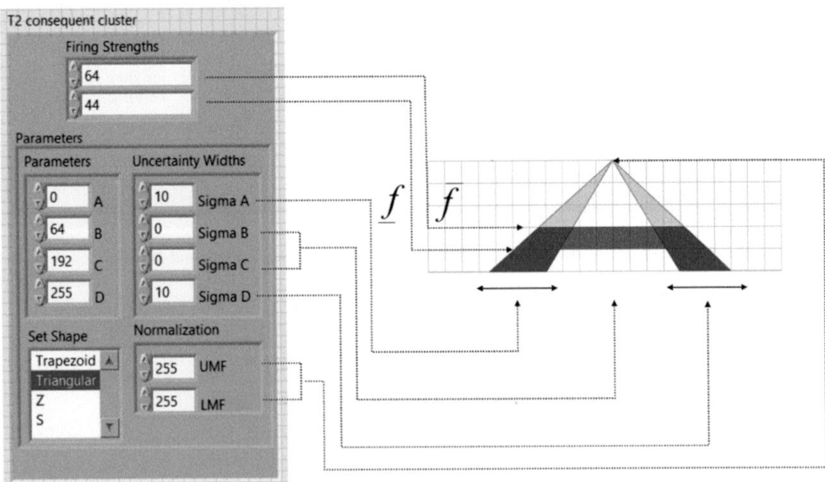

Fig. 4.12 Parameter meanings for T1 and T2 consequent sets

A, B, C, and D are the characteristic points for the trapezoidal MF. These same elements can describe a triangular MF, so the characteristic point B can be supplied for B and C, i.e., just set $B = C$ equally for selecting the B parameters of the triangle. Set A and B if you want to use the S-shape MF leaving the parameters C and D equal to zero. This is the same for the Z-shape MF; so you must set the parameters C and D and leave in zero A and D.
- Uncertainty: This is the same for the uncertainty values if you are using T2FLS. If you elect the trapezoidal MF, please set Sigma A, Sigma B, Sigma C, and Sigma D; if you select the triangular MF, set Sigma A and Sigma D; if you select the Z-shape MF, set Sigma C and Sigma D; and if you select the S-shape MF, set Sigma A and Sigma B
- Normalization: If you are using T1FLS, there is only a normalization value which defines the set height. For T2FLS, there are two normalization values clustered: one for UMF and the other for LMF.

2. Its corresponding firing strength.

- The T1 firing strengths: For T1FLS, the firing strength is the membership value obtained from the inference step. This defines an α—cut where the consequent set will be trimmed.
- The T2 firing strengths: For T2FLS, the firing strength is a cluster of two membership values obtained from the inference step also. This provides two α—cuts where the consequent set will be trimmed in UMF and LMF.

– Describes a single consequent set.
– It is equal to the other elements in the array.

This way, a T1/T2 inferred set functional block must be connected as you can see in the red dotted lines in both Figs. 4.10 and 4.11. Notice that each consequent set is an element of the resulting array (Build Array).

A configuration cluster provides to the T1 and T2IS the necessary information for building the consequent T1 or T2FS, i.e., how much consequent sets are considered, their shape, distribution, uncertainty (if T2) and height (s). Depending on the number of consequent sets in the system, there will be that number of firing strengths. Each configuration cluster and firing strength must be clustered for each consequent set; also every resulting cluster must be set into an array. So, only one array is needed to process the IS.

Both examples use an IRQ block and a flat sequence that helps the host computer synchronize with the target FPGA execution. For additional information about the target synchronization, please refer to Chap. 2.

In hardware terms, searching for the IS requires of a memory whose address works as the domain (each output discourse universe value represents the memory location) and the memory content works as the codomain (the IS membership values are the memory data contained in that location). The FPGA fuzzy logic chapter uses a single VI-defined Memory (LUT) for the T1 inferred set and; two VI-defined Memories (LUT) for the T2 inferred set. The generalized centroid is calculated from these memories.

In some cases, more than one memory is needed, especially when the system is considered as MIMO or when using T2FLS. Each output should use a memory. The memory must contain 256 locations or addresses and each location must store unsigned 8-bit fixed point values. Each location must be addressed from 0 to 255.

The output of a T1 inferred set functional block, is a memory reference. But, in the case of a T2 inferred set functional block, its output is a cluster of two memory references. Depending on the type of fuzzy system used, those memory references can be used to defuzzify and calculate the crisp output.

The next Chapter presents several resources for translating a fuzzy set into a crisp value, both for T1FLS and T2FLS.

Examples

This chapter includes two examples about T1 and T2 inference machines, which are located in the examples\chapter4 directory.

Those VI contains a T1 and T2 fuzzification stage whose membership degrees are used as inputs in their corresponding inference machines.

All you have to do is

1. Create a new FPGA project and select your appropriated FPGA target.
2. Add to the project these files

    (a) T1 rule set example VI: This VI helps the user to verify the T1 interference performance with an interactive front panel which includes its T1 fuzzification stage (see Fig. 4.13).
    (b) T2 rule set example.VI: This VI help the user verify the T2 inference performance with an interactive front panel which includes also its T2 fuzzification stage (see Fig. 4.14).

3. Compile both VIs.
4. Run them separately with their appropriated set parameters.

Observe the differences between both Vis and notice that a T2 inference machine can be implemented with two T1 inference machines.

Also, two VI examples are provided for a T1 inferred set and a T2 inferred set calculation. These examples let you interact with all the parameters of the sets and see the response in the corresponding indicators.

1. In the same new project, add the following VI files:

    - T1 inferred set example.vi: This VI fills the target memory MF in order to calculate the T1 inferred set.
    - T2 inferred set example.vi: Similar to the last VI, this VI fills both target memory UMF and LMF in order to calculate the T2 inferred set.
    - Graph a T1 inferred set example.vi: This VI let the user interact with the consequent Type-1 inferred set parameters and normalization values, and all

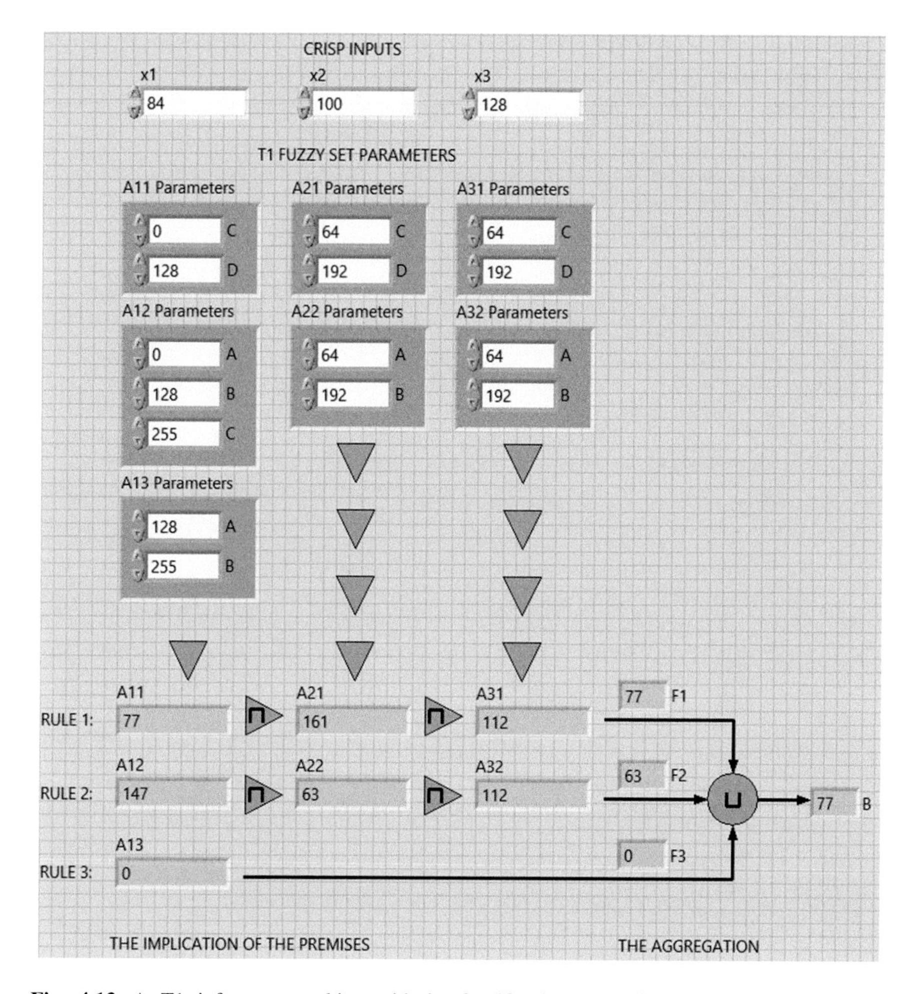

Fig. 4.13 A T1 inference machine with its fuzzification stage included. Notice that each membership value is used in the inference machine for consequent selection

the available shapes in the chapter. The user can verify graphically by means of a waveform chart its corresponding shape whenever the firing strength is changed (see Fig. 4.15).

- Graph a T2 inferred set example.vi: This VI let the user interact with the consequent Type-2 inferred set parameters and normalization values and all the available shapes in the chapter. The user can verify graphically by means of a waveform chart its corresponding shape whenever the firing strength is changed (see Fig. 4.16).

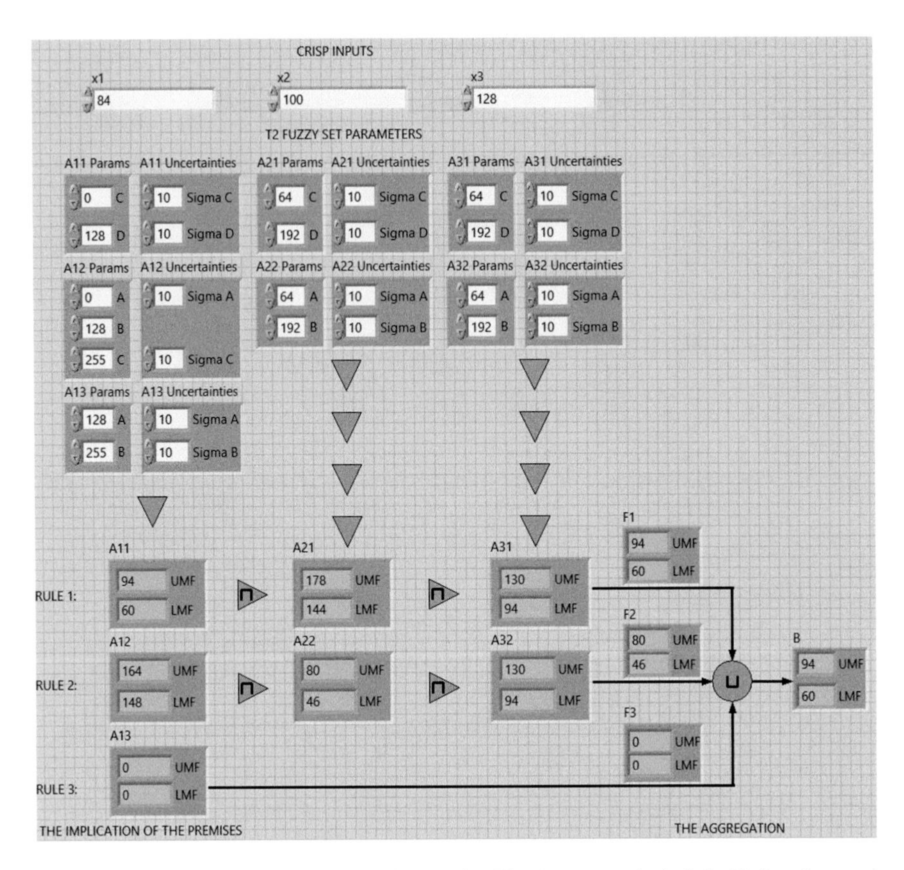

Fig. 4.14 A T2 inference machine with its fuzzification stage included. Notice that each membership value is clustered and each rule association follows this clustering

2. Add three FIFOs in the FPGA target named MF, UMF, and LMF with the following characteristics:
   - Target-Host DMA FIFO,
   - Unsigned 8-bit fixed point data length, 8-bit for the integer part,
   - Do not change the memory access arbitration.
3. Compile the following Vis in the current target
   - T1 inferred set example.vi
   - T2 inferred set example.vi
4. Now, you can execute and interact with the T1/T2 inferred sets and see how their shapes change whenever you change the parameter and firing strength controls.

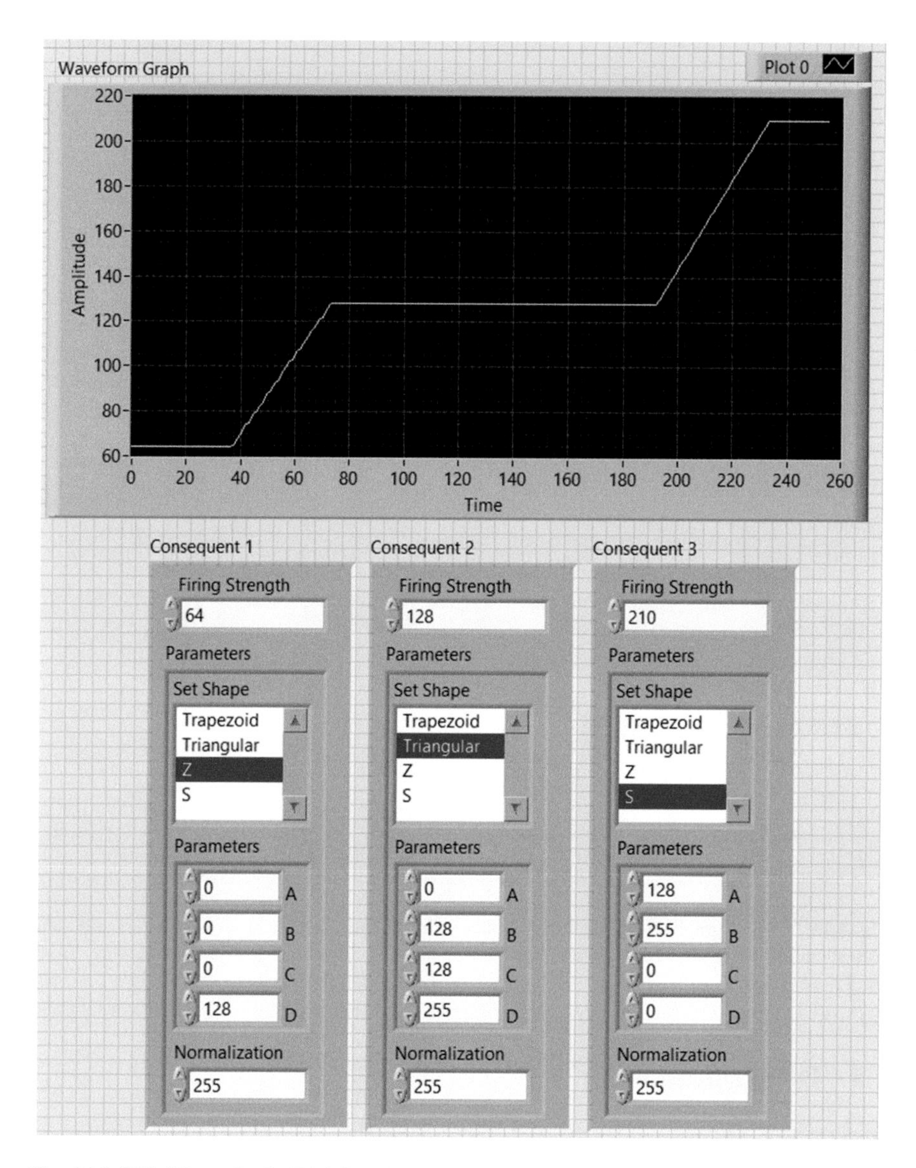

Fig. 4.15 This VI graphs the T1 inferred set according to the consequent set parameters and their corresponding firing strengths

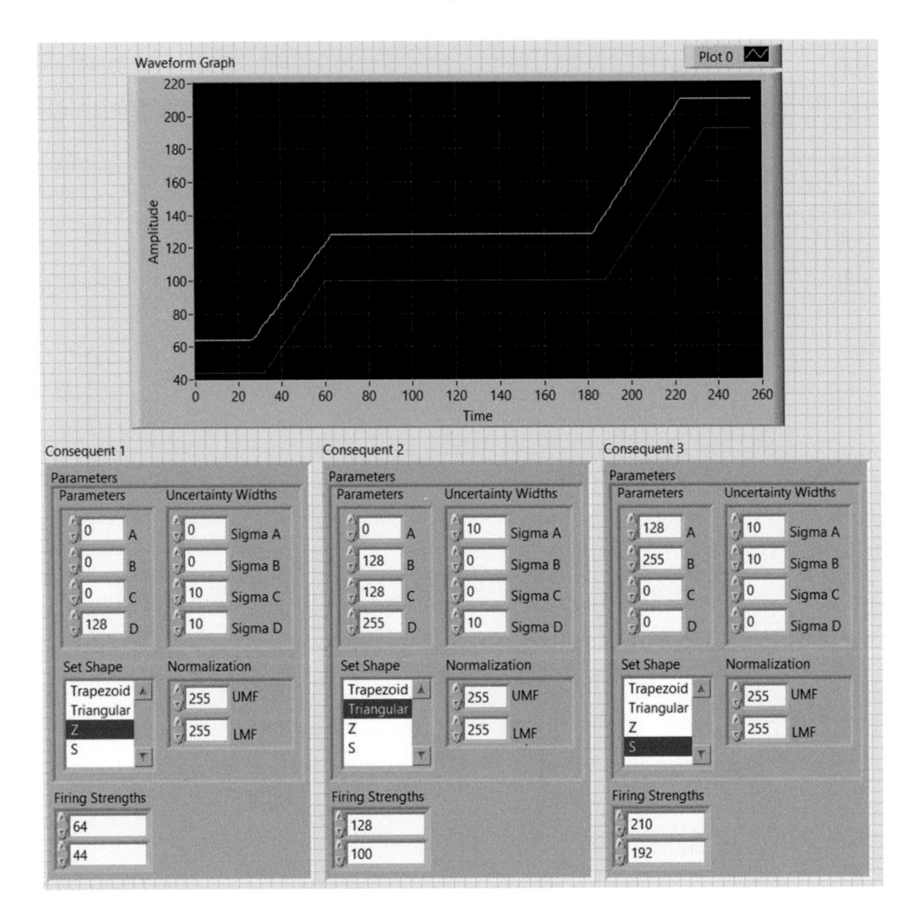

Fig. 4.16 This VI graphs the T2 inferred set according to the consequent set parameters and their corresponding firing strengths. Notice that their parameters firing strengths and shape differ from the T1

4.5 Defuzzification

4.5.1 T1 Mamdani Model the Centroid

As it was explained in chapter one, the T1 inferred set VI is comprised of a LUT memory which is filled with the membership values according to the firing strengths and the FS shape. That memory is defined in VI, so that it is not necessary to define that in the project. The output of that VI is a memory reference, which is used in the following VI: the Centroid VI. The Centroid VI uses this memory reference for reading the membership values and calculating the centroid. If the memory is empty, then the centroid calculation must be 128. The T1 Mamdani model do not require additional information for calculating the crisp output. All you have to do is

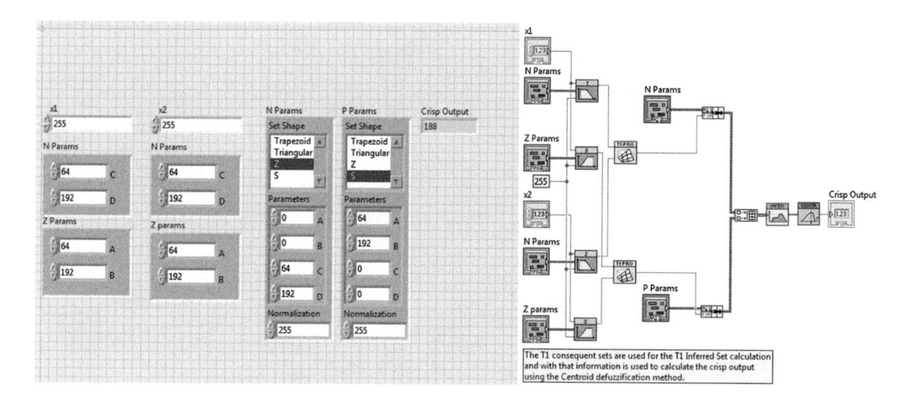

Fig. 4.17 A T1 Mamdani model using the T1 Inferred set and the Centroid VIs

to connect the T1 inferred set VI indicator to the Centroid VI control. No matter how many consequent FSs exists in the variable. If you want to add several output variables, you will need a T1 inferred set—Centroid pair for each output variable (Fig. 4.17).

4.5.2 T2 Mamdani Model the Karnik–Mendel Algorithm

The KM VI is designed for the T2FLS defuzzification. In the literature, it is considered as a Type-Reduction method called as the Karnik–Mendel Algorithm. Two versions are available: the Non-Iterative KM Algorithm and the Iterative KM Algorithm.

Their differences lie in the number of iterations the centroid search is performed. In this case, the use of the Iterative KM Algorithm is recommended.

Alike the T1 Mamdani model (as explained in Chap. 1), the T2 inferred set VI is comprised of two LUT memories, one for the UMF and another for the LMF. Each memory contains every membership value that is stored in their corresponding memory. The output of this VI is a cluster of two memory references.

Once the T2 inferred set is ready, these memory references are connected to the defuzzification and type-reduction method for calculating the crisp output. The KM VI has an input cluster that contains the VI-defined memory reference. Each memory element is used to calculate both left and right centroids.

Alike the T1 Mamdani model, the T2 Mamdani model also do not require additional information for calculating the crisp output. You just have to set the appropriated parameters for calculating the T2 inferred set, select the consequent T2FS shape, and connect the T2 inferred set output indicator to the KM input control (see Fig. 4.18).

If you want to add several output variables, you will need a T2 inferred set—KM pair for each output variable.

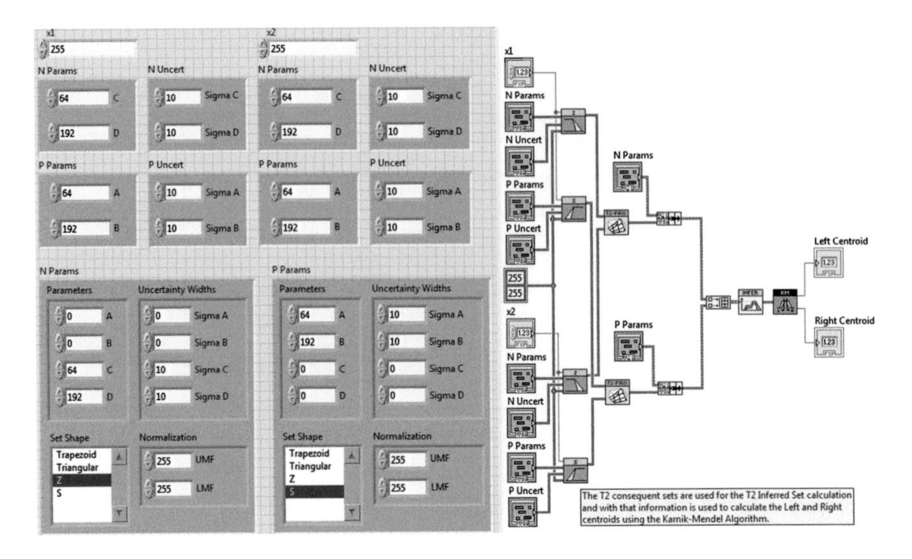

Fig. 4.18 A T2 Mamdani model using the T2 inferred set and the KM Vis

4.5.3 The Enhanced Karnik–Mendel Algorithm

This is the best version of the KM Algorithm, better known as EKM, where the centroid search is completely optimized. The authors claim that this algorithm surpasses the original algorithm over 50 %.

The use of the EKM is widely recommended where the time restrictions are strict. Due to its centroid time reduction, this works better than the Non-Iterative and Iterative KM Algorithms.

The way you connect this VI is exactly the same as the other two versions.

4.5.4 The Nie–Tan Method

The NT VI is designed also for the T2FLS defuzzification. In the literature, it is considered as a defuzzification method called as the Nie–Tan Method.

This method accomplishes the type-reduction considering a single embedded set. This set is located in the middle of both UMF and LMF. After this set is calculated using

$$\theta_{\mathrm{NT}} = \frac{1}{2}\left(\bar{\mu} + \underline{\mu}\right)$$

the final output is only a single centroid of that FS. This method is just an approximation and do not require complex calculations.

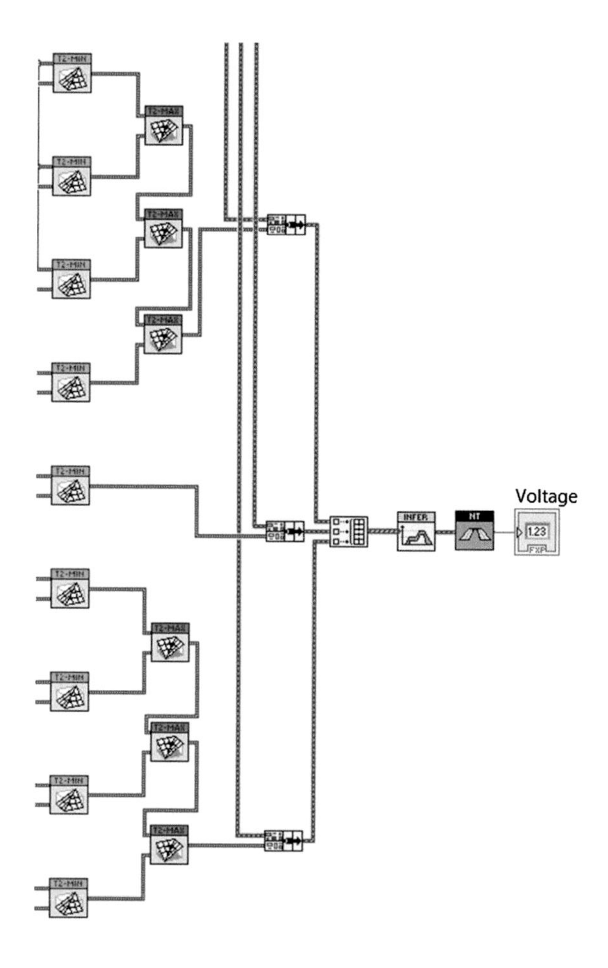

Fig. 4.19 Replacing the KM algorithm for the NT method. Notice that you do not require average the left and right centroids due to this method obtains a single output

The NT Method.vi is also available in this chapter. You just have to replace the KM Algorithm.vi in Fig. 4.18 for the Nie–Tan Method.vi as can be seen in Fig. 4.19.

This method represents the fastest option for Mamdani T2 defuzzification.

4.5.5 *The Takagi–Sugeno Model*

The Takagi-Sugeno-Kang model, also known as the TSK model, does not use the FS shape.

As stated in literature, the TSK model is based also on the rule set. Its knowledge base decides how its behavior will be.

For both T1 and T2FLS, the firing strengths are used. This model differs from the Mamdani model in the sense of how the consequent FS is used. The consequent

FS in the TSK model is a polynomial. Each polynomial corresponds to a consequence.

For both T1 and T2FLS, you can add up to 64 consequent sets or polynomials and 16 input variables, as stated in Appendix 2, Section *Polynomials*. Each polynomial can be of second order at most. Alike the T1 and T2 inferred set, you can add several output variables adding several T1 TSK or T2 TSK VIs, which are available also in the toolkit.

After the inference part, you will have several inferred values, which are the heights of the inferred set when the Mamdani model is used. These values weight the way each polynomial will contribute for the crisp output.

With the TSK model, it is not necessary to use the T1 or T2 inferred set VI; however, you must take into account some considerations.

For both T1 TSK and T2 TSK VI, you must prepare an array which must include the firing strength values and an array which must include their corresponding input variable and a LUT memory, which can be a project-defined memory or a VI-defined memory.

That memory must contain the coefficients of each polynomial

$$P_i{:}\left\{[a,b,c]_{\forall x_j}, x_j\right\}{;}f_i$$

where $P_i{:}a_j x_j^2 + b_j x_j + c_j$, and x_j is the j-th input variable and $[a,b,c]_j$ is the j-th coefficient set for its corresponding polynomial, $i = 1,2,\ldots M$ and $j = 1,2,\ldots m$.

Due to hardware restrictions, $M = 64$ and $m = 16$.

Note that each element i in the array selects the i-th firing strength. Please refer to Appendix 2, Section Polynomials.

The polynomial coefficient is stored into each memory location. Due to the reason that there can be $3 \times m \times M = 3072$ coefficients, you can create up to 64 polynomials for each firing strength. For instance, if you want to represent the polynomial $2x_1 + x_2 + 3$, you can represent the first input variable in the polynomial $2x_1 + 3$ with the coefficients $\{0,2,3\}$ and the rest of the polynomial x_2 with the coefficients $\{0,1,0\}$ (see Fig. 4.20).

After the polynomial value is calculated, this value is multiplied by its corresponding firing strength f_i.

Sometimes, when the TSK model is used, the aggregation part is not included in the inference step, although they are not necessarily dispensable.

As you can see in Fig. 4.21, the polynomial coefficients are not visible because they are into the LUT memory for the T1 TSK VI. Note that the firing strength and the input variable are numerical control in two separated arrays. Also, for accessing the coefficient of x_2 you may modify the location 193. The rest of the coefficients must be zero if you do not want them to affect the final result.

As you can see in Fig. 4.22, the polynomial coefficients for the T2 TSK VI are also configured inside the corresponding LUT memory.

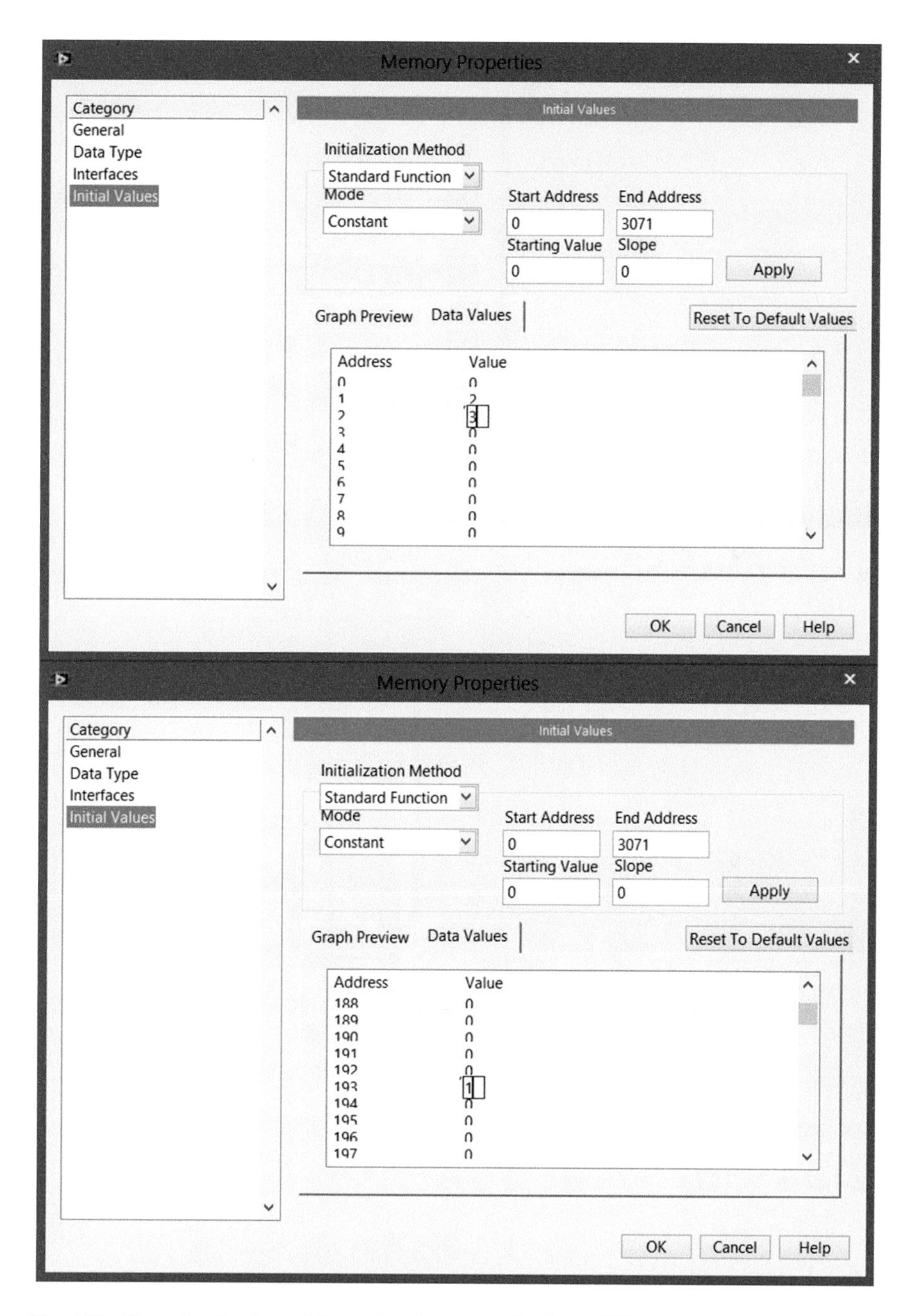

Fig. 4.20 Example of polynomial creation: $2x_1 + x_2 + 3$ using a LUT memory. You can type the coefficient values manually or you can use the Polynomial Generator.vi

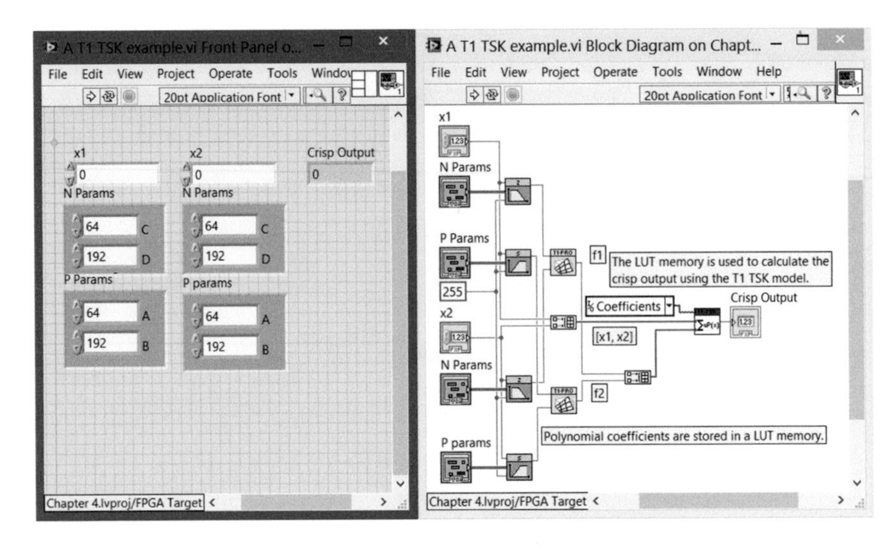

Fig. 4.21 A T1 TSK model example

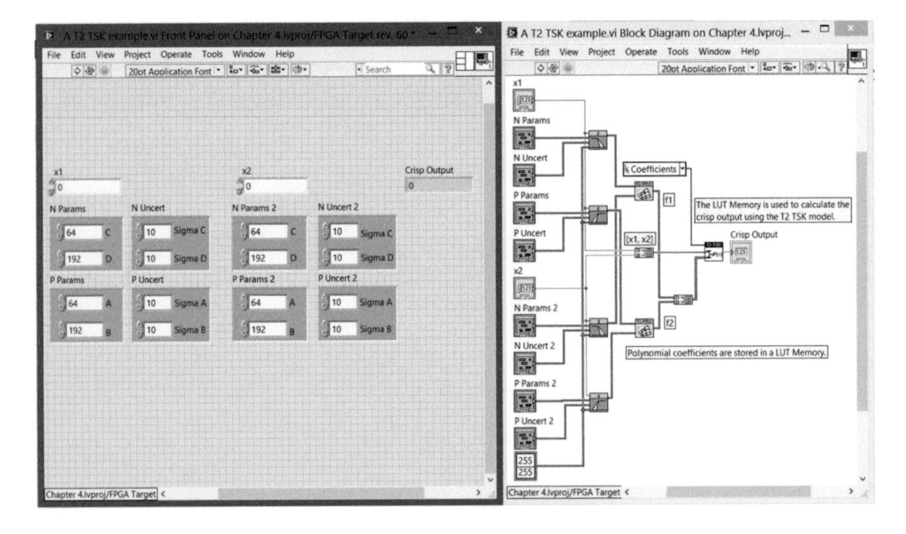

Fig. 4.22 A T2 TSK model example

4.6 Examples

The FPGA Fuzzy Logic Chapter includes a defuzzification models VI example for each T1FLS and T2FLS located in the examples\chapter4 directory. These examples let you interact with several parameters just as the firing strengths (for the

Mamdani model) and the polynomial coefficients (for the TSK model) and see how the crisp output is moved along its corresponding output discourse universe.

5. Open a new project and add the following VI files:

   - A T1 Mamdani example.vi: This VI let you set the firing strength values, just as the inference part does it, but you can set them manually. So, you can see how the crisp output changes as long as the firing strengths are changed (see Fig. 4.17).
   - A T2 Mamdani example.vi: Similar to the last VI, this VI let you set the firing strength values both for UMF and LMF, just as the inference part does it, but you can set them manually. So, you can see how the crisp output changes as long as the firing strengths are changed (see Fig. 4.18).
   - A T1 TSK example.vi and A T2 TSK example.vi: These VIs provide you several controls which let you set the crisp input values, the firing strength values and their corresponding consequent polynomials (see Figs. 4.21 and 4.22).

6. Add a LUT Memory to the project or insert a VI-defined memory for the coefficients.
7. Compile the recently added VIs in the current target.
8. Now, you can execute them separately and interact with the Mamdani/TSK parameters and see how the crisp output moves along its output discourse universe.

4.7 Study Cases

This section is dedicated to the FPGA fuzzy logic implementation. Several FPGA implementations were proposed and proved to be feasible for T1FLS and T2FLS applications.

Two applications were proposed: a DC servomotor and an electric wheelchair. Each application is proved with a T1FLS and is demonstrated its efficiency. Also, for the T2FLS, only the DC servomotor is proved.

At the end of this section, you will understand the advantages and differences between the T1FLS and T2FLS when they are implemented in FPGAs.

4.7.1 T1FLS Validation

DC Servomotor

Preliminaries

For the servomotor control, you will need several components, modules, and the FPGA fuzzy logic.

In this example, the FPGA target used for demonstrating the performance of the program is a CompactRIO NI cRIO-9014. The NI-9014 has a processor that manages the communications between a PC and the FPGA device. This communication is performed by Ethernet. This Ethernet communication is used for programming, data exchange, and debugging. In a cRIO, you can connect several modules which give the ability of processing analog–digital information, and several additional characteristics. Please visit the NI webpage for additional information about this cRIO model.

On the other hand, we do not have detailed information about the DC servomotor dynamics; the only information we know is the following:

- Supply voltage: 0 5 V
- 0 *rads* position voltage: 0 V
- π *rads* position voltage: 5 V

For the position measurement, an encoder was used; its output was driven to a DAC. Its analog output was driven to a NI-9201. The NI 9201 is an input module for processing analog information with 8-Ch, which can accept voltages up to ±10 V, process the information with a velocity of 500 kS/s; provides a precision of 12-bit. This module belongs to the C-Series. We will use just a single input channel of this module.

Due to the position of the servomotor, rotor can be controlled by the voltage applied to its terminals; the fuzzy controller must provide the necessary voltage in order to obtain the desired position. For this purpose, a NI-9263 C-Series module is used. The NI-9263 module has four channels, can process information to a velocity of 100 kS/s, can process data information of 16-bit, and provide up to ±10 analog voltage. For our purpose, we will use a single output channel.

Additional information about this and another C-series modules can be found at http:\\NI.com.

So, in resume, we will need

- NI cRIO-9014,
- NI-9201,
- NI-9263, and
- DC servomotor.

For this application, all the electronic circuitry around the servo was developed. Figure 4.23 shows a picture of the system and Fig. 4.24 shows all the circuitry and their connections with the target cRIO.

4.7.1.1 Implementation

The fuzzy logic system that will work as the controller is now proposed. With this limited information, we can control the servomotor.

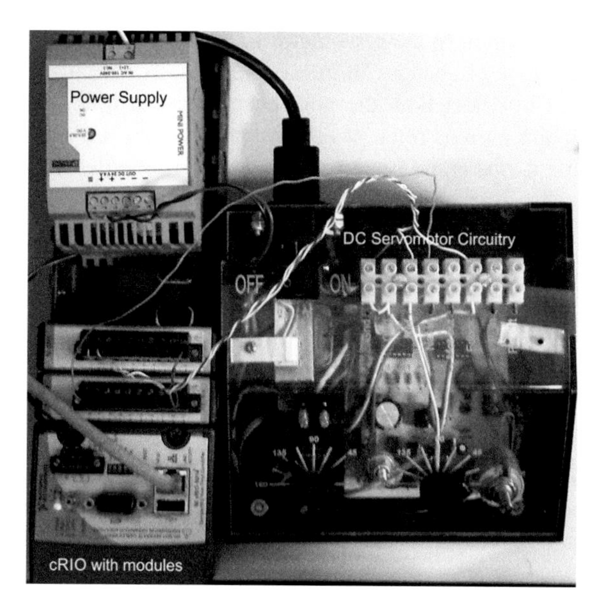

Fig. 4.23 An illustration of the servomotor circuitry and their connections with the cRIO

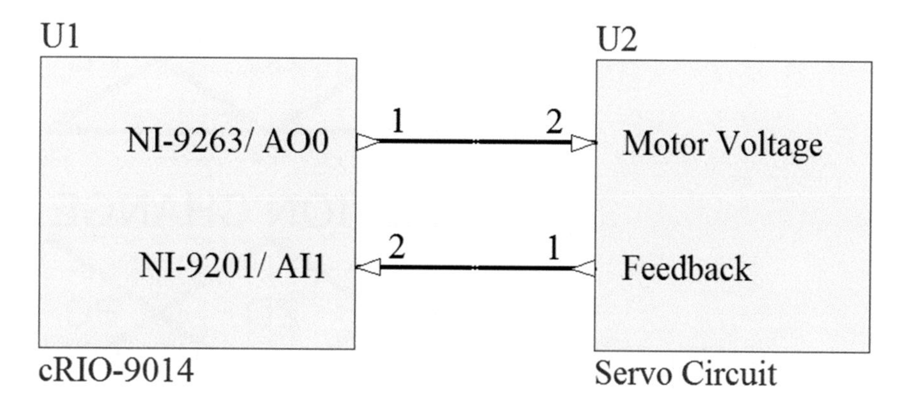

Fig. 4.24 The connection diagram of the servomotor circuitry and the cRIO

The following steps were followed to design the T1FLS:

1. Define the number of inputs. In this case, we need the current position voltage, as stated in the last section, for calculating the *Position Error*

$$\theta_e = \theta_{ref} - \theta$$

where θ_e is the position error, θ_{ref} is the reference position, and θ is the current position.

Every time the position of the servomotor rotor changes, the voltage in terminal FeedBack in Fig. 4.24 changes in interval [0, 5] volts.
Also, we need to detect how the position changes in time; this is useful for calculating if the current rotor position approaches to the reference. For this variable, name it *Position Change*, we can perform the derivative of the position. A derivative of position can be calculated simply as

$$\frac{d\theta}{dt} = \theta(t) - \theta(t-1)$$

The output variable for controlling the servo is the applied voltage, but we want to move the servo rotor to the left or right in order to get the desired position. Evidently, these linguistic variables must be related to a voltage; so the output variable is called *Movement*.
The set distribution can be seen in Fig. 4.25.

2. Define the number of T1FS per variable. Now that we defined the variables, we need to think about the possible linguistic variables for each T1FS.

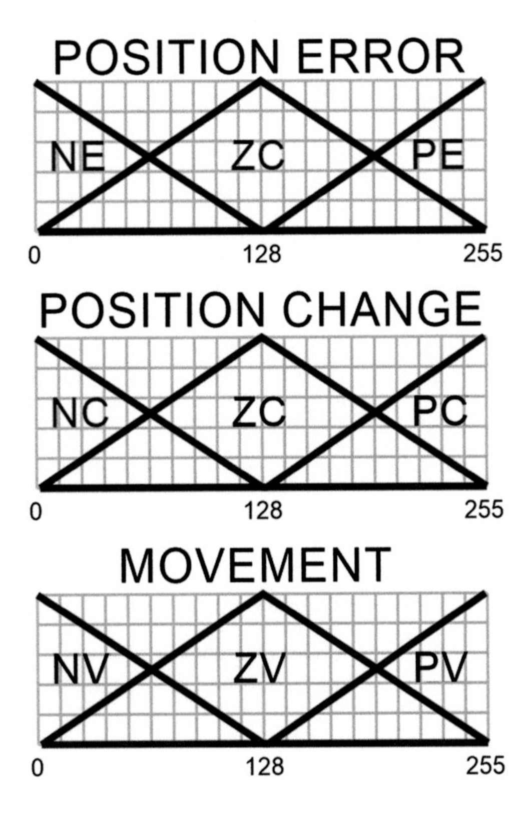

Fig. 4.25 T1FS distribution for the servomotor control

For instance, the *Position Error* (e) may have three states

(a) When the error i is below the reference, i.e., a *Positive Error* or;
(b) When the error is exactly in the reference, i.e., a *Zero Error* or;
(c) When the error is over the reference, i.e., a *Negative Error*.

These linguistic variables are associated to three T1FS, characterized by a T1MF.
Now, we can use the FPGA fuzzy logic program fuzzification tools: Due to the positive and negative error tags are the left and right end sets, we can use the T1 Z-Shape MF and T1 S-Shape MF, correspondingly. For the zero error, we can use the triangular MF.
For the variable *Position Change* (c), due to the differential is also a subtraction of two values, we may have also three states

(a) When the rotor movement is increasing in a direction, i.e., a *Positive Change* or;
(b) When the rotor movement is not changing, i.e., a *Zero Change* or;
(c) When the rotor movement is decreasing in the contrary direction, i.e., a *Negative Change*.

As proposed for the *Position Error* variable, the left and right end sets are related to the positive and negative changes, so we can use the T1 Z-Shape MF and T1 S-Shape MF, correspondingly. For the zero change, we can use the Triangular MF also.
For the output variable *Movement* (M), we may have also three states

(a) When you want to move the rotor to the left, i.e., a *Left Movement* or;
(b) When you do not want to move the rotor, i.e., a *Null Movement* or;
(c) When you want to move the rotor to the right, i.e., a *Right Movement*.

3. After we established that we have three T1FS per variable, we may have up to nine rules in the rule set. For this purpose, we proposed the rule set presented below. This rule set can be built in the FPGA.

r_1	$e:\text{NE} \sqcap c:\text{NC} \Rightarrow v:\text{NV}$
r_2	$e:\text{NE} \sqcap c:\text{ZC} \Rightarrow v:\text{NV}$
r_3	$e:\text{NE} \sqcap c:\text{PC} \Rightarrow v:\text{NV}$
r_4	$e:\text{ZE} \sqcap c:\text{NC} \Rightarrow v:\text{NV}$
r_5	$e:\text{ZE} \sqcap c:\text{ZC} \Rightarrow v:\text{ZV}$
r_6	$e:\text{ZE} \sqcap c:\text{PC} \Rightarrow v:\text{PV}$
r_7	$e:\text{PE} \sqcap c:\text{NF} \Rightarrow v:\text{PV}$
r_8	$e:\text{PE} \sqcap c:\text{ZF} \Rightarrow v:\text{PV}$
r_9	$e:\text{PE} \sqcap c:\text{PF} \Rightarrow v:\text{PV}$

4. Now that all the variables, sets, and rules are defined, we proceed to calculate the crisp output based on the *firing strengths* f_i, the *T1 inferred set* the consequent T1FS *shape* using the *Centroid* defuzzification method.

Finally, the T1FLS for the servomotor is designed as you can see in Fig. 4.26 All those functional blocks are available in the FPGA fuzzy logic program.

After the T1FLS is built in the target FPGA device, using the FPGA fuzzy logic program, you can create the control system where you can connect all the variables with analog inputs and outputs.

You can also use some *Magnitude Converters* (see Chap. 1 or refer to Appendix 2 for the utility functions) for adjusting the scale and offset of the numerical values. Figure 4.27 presents the control system where the T1FLS is implemented.

4.7.2 Electric Wheelchair

Preliminaries

For the electric wheelchair, you will need also several components. We used a wheelchair Quickie model P222-SE. Also, the joystick that provides the user the desired movement and direction was substituted by the NI-9263; so we can use a T1FLS for controlling its forward–backward movement, left–right steering and speed.

The wheelchair has two motors that provide the user the desired movement and steering. So, if both motors move ahead at same velocity, the wheelchair will move to the front. If one of the motors decreases its velocity, then the wheelchair will steer to one side or another.
Now, we have information about the wheelchair also

- Operating voltage: 4.82 6.82 V
- Full-speed forward movement and right steering voltages: 6.82 V
- Full-speed backward movement and left steering voltages: 4.82 V
- Null movement and steering voltages: 5.82

We will use the same target FPGA, the cRIO-9014. Also, we will need the two analog output channels from the NI-9263 for providing the specific voltages to both motors.

The purpose of this T1FLS is to make the wheelchair move along unstructured environments without collisions. Also, it should be able to void static and dynamic obstacles. For this purpose, we will use three ultrasonic distance sensors, specifically the Parallax PING))) sensors.

The Parallax PING))) sensor sends an ultrasonic burst to the environment and whenever the signal bounces in the obstacles, the burst comes back to the sensor;

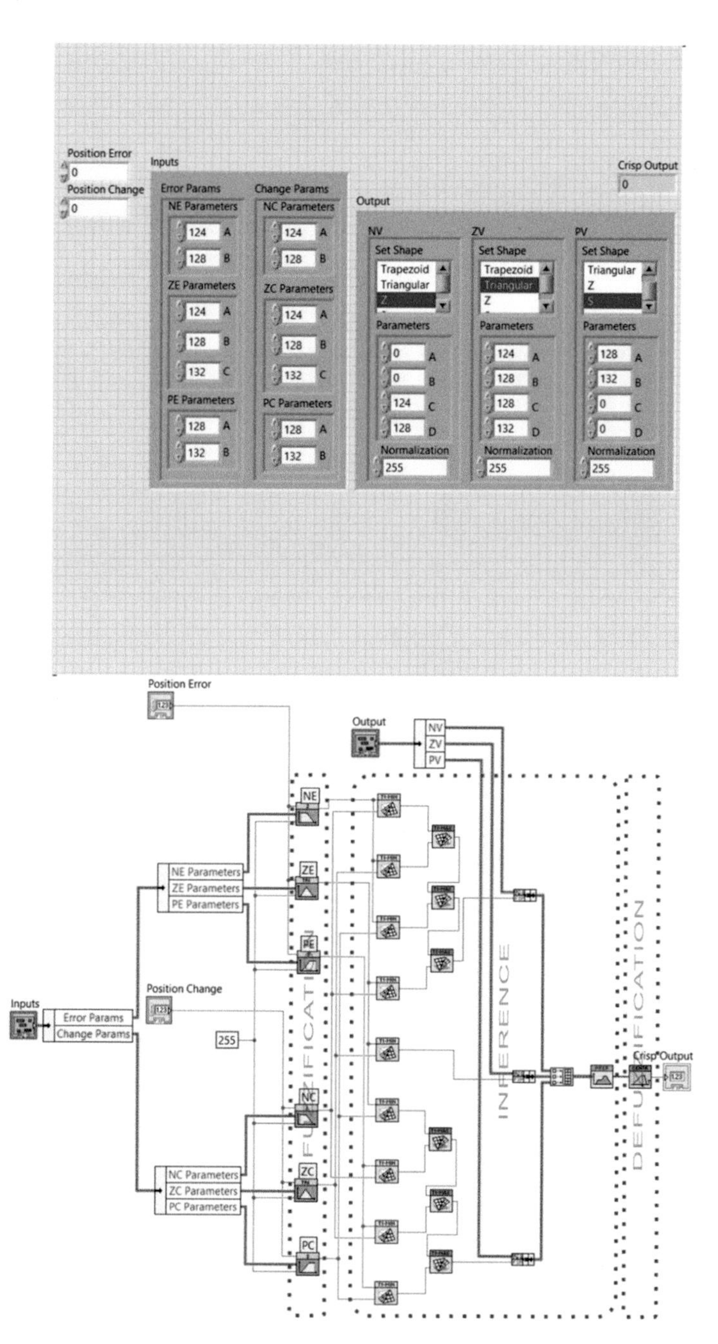

Fig. 4.26 The T1FLS for the servomotor control

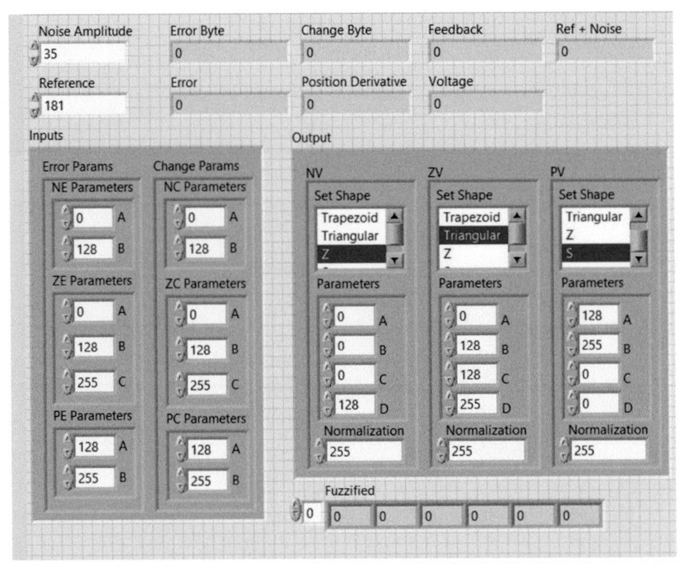

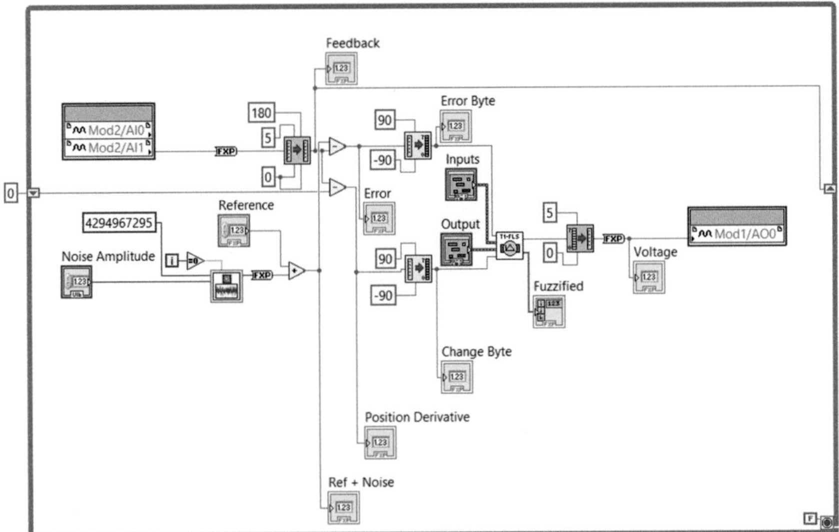

Fig. 4.27 The T1FL control system for the servomotor

then the sensor retrieves a digital pulse that represents the time the burst elapsed in coming back to the sensor. Due to (70)

$$d = ct$$

where c is the sound speed and t is the time in seconds.

After calculations, the pulse is translated into distance. Because of this, we will need digital inputs and outputs for calculating the distance of each sensor to the obstacles. The program provides a VI called Parallax Ping))) Sensor Decoder.vi which performs this translation. You only have to define the clock source, define the input and outputs DIOs.

The information about the sensors is the following:

- Maximum distance measurement: 373 cm approximately (Fig. 4.28).
- There are three sensors: the left sensor s_1, the right sensor s_2, and the back sensor s_3. Both side sensors are not exactly oriented perpendicularly. They are located in both frontal corners (diagonally) of the wheelchair, so additional information of the frontal obstacles are considered (see Fig. 4.29).

For the sensor measurement, the NI-9401was used. The NI-9401 is another C-series module that provides eight 5 V/TTL High-Speed Digital I/Os. The PING)))

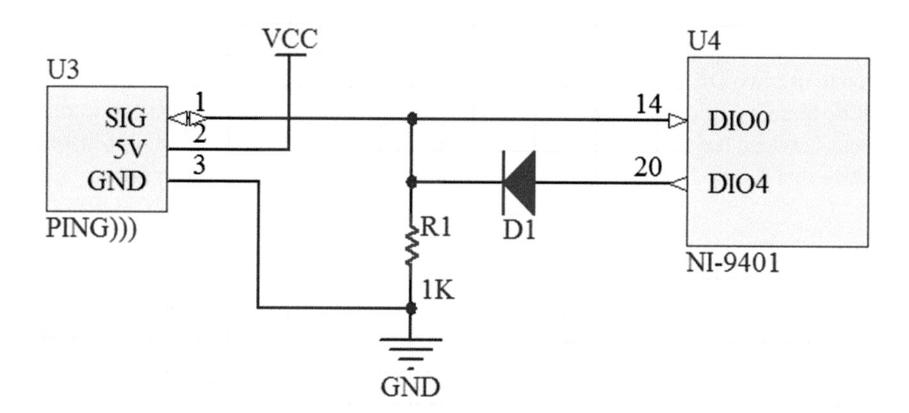

Fig. 4.28 Bidirectional port to input–output ports conversion for the PING))) ultrasonic sensor

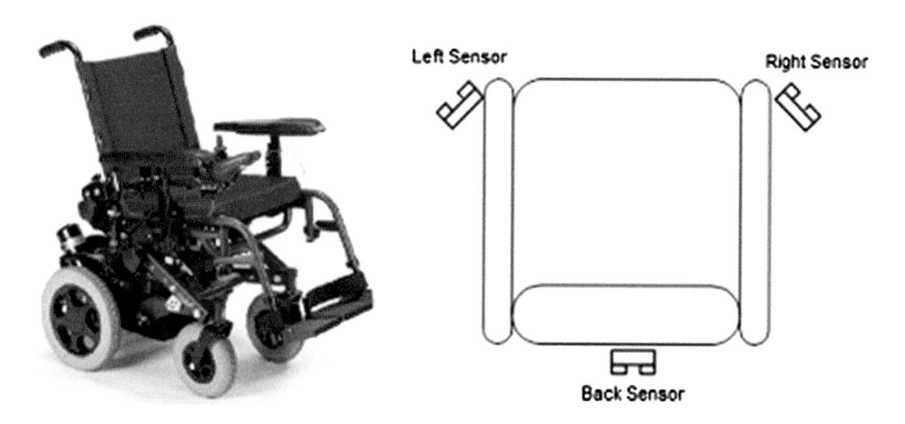

Fig. 4.29 The wheelchair and their ultrasonic sensors

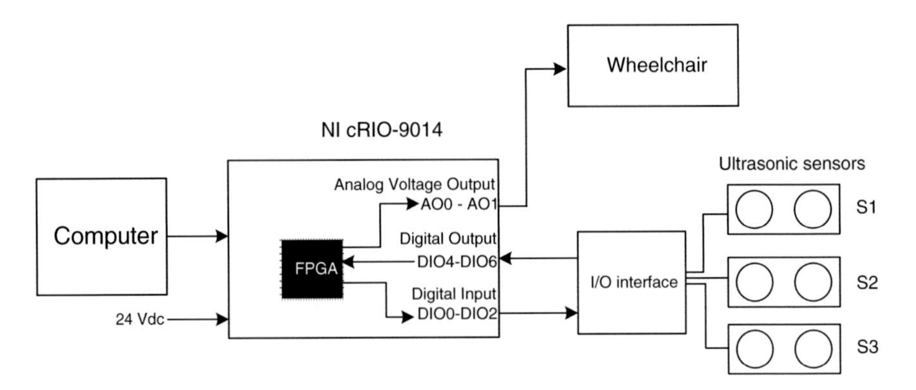

Fig. 4.30 Connection diagram: the wheelchair, the target FPGA and the ultrasonic sensors

sensor has a single bidirectional port that triggers the measurement (sending the ultrasonic burst) and at the same time, retrieves the resulting pulse.

Due to the reason that NI-9401 ports cannot be configured as bidirectional, we will need to use two DIOs, one as input and another for output per sensor. So, we need six DIOs, three as inputs and three as outputs. Figure 4.30, shows the connection diagrams between the wheelchair, the target FPGA (crio-9014) and the PING))) sensors. Notice that DIO3–DIO0 are configured as inputs and DIO7–DIO4 as outputs.

4.7.2.1 Implementation

Now, we must define the T1FLS for the wheelchair application. The following steps were followed to design it:

1. Define the number of inputs. All the information that we have in this moment are the three distance measurements obtained from each PING))) sensor. So, we have to think about what and how we want the chair to behave in the non-structured environment.
   Prior to defining new variables, let us state the variables s_1 as the *Left Side Sensor*, s_2 as the *Right Side Sensor*, and s_3 as the *Back Side Sensor*.
   As we do not have a reference to follow, since this has no feedback (because the wheelchair does not know its position in the environment), we need to make the wheelchair follow something. Because of this, let s be the *Middle Trajectory* of the wheelchair; it means that whenever the wheelchair goes ahead it should force the distance measurements of each s_1 and s_2 as they equal all the time. This can be done if

$$s = s_1 - s_2 = 0$$

This way, the wheelchair can pass through corridors trying to keep the same distance in each side.

Also, we need the wheelchair to void dynamical obstacles whenever they cross in its trajectory. If we use the distance measurements from each side sensor, we could know how fast or slow a dynamical obstacle approaches to the wheelchair; this way the wheelchair could decide how to void it. This can be performed similarly as the servomotor with

$$\frac{\mathrm{d}s_1}{\mathrm{d}t} = s_1(t) - s_1(t-1)$$
$$\frac{\mathrm{d}s_2}{\mathrm{d}t} = s_2(t) - s_2(t-1)$$

So, the variable names for these are *Left Approaching* and *Right Approaching*, respectively.

Finally, we have six input variables.

The output variables, as stated in previous paragraphs, are the Movement (*M*) and the Steering (*D*).

The set distribution can be seen in Fig. 4.31.

2. Define the number of T1FS per variable. For the variables s_1, s_2, and s_3, two sets are enough for determining if the obstacle is near or not; so the linguistic variables are *Near* (*N*) and *Far* (*F*). For $\mathrm{d}s_1/\mathrm{d}t$ and $\mathrm{d}s_2/\mathrm{d}t$ we need also two linguistic variables: *Getting Slow* (GS) and *Getting Fast* (GF). For all these variables, their T1FS can be represented with the T1 Z-Shape MF and T1 S-Shape MF. For the variable s, we need three linguistic variables, such as Left (*L*), Null (*N*), and Right (*R*), which can be represented with a T1 Z-Shape MF, a T1 triangular MF, and a T1 S-Shape MF, respectively.
3. Then, we propose the rule set for this application. The total number of possible rules are $2 \times 2 \times 2 \times 2 \times 2 \times 3 = 96$. Due to the derivative of each s_1 and s_2 works to detect when an obstacle is approaching fast, we just use the GF set of each variable. The proposed rule set for this application is presented below.

r_1	$s_1 : C \sqcap s_2 : C \sqcap s_3 : C \Rightarrow M : N \sqcap D : N$
r_2	$s : N \sqcap s_1 : F \sqcap s_2 : C \Rightarrow M : \mathrm{MF} \sqcap D : L$
r_3	$s : P \sqcap s_1 : C \sqcap s_2 : F \Rightarrow M : \mathrm{MF} \sqcap D : R$
r_4	$s : N \sqcap s_1 : C \sqcap s_2 : C \Rightarrow M : B \sqcap D : R$
r_5	$s : Z \sqcap s_1 : C \sqcap s_2 : C \Rightarrow M : B \sqcap D : N$
r_6	$s : P \sqcap s_1 : C \sqcap s_2 : C \Rightarrow M : B \sqcap D : L$
r_7	$s : N \sqcap s_1 : F \sqcap s_2 : F \Rightarrow M : F \sqcap D : \mathrm{ML}$
r_8	$s : Z \sqcap s_1 : F \sqcap s_2 : F \Rightarrow M : F \sqcap D : N$
r_9	$s : P \sqcap s : F \sqcap s_2 : F \Rightarrow M : F \sqcap D : \mathrm{MR}$
r_{10}	$\mathrm{d}s_1 : \mathrm{GF} \sqcup \mathrm{d}s_2 : \mathrm{GF} \Rightarrow M : \mathrm{MF} \sqcap D : N$

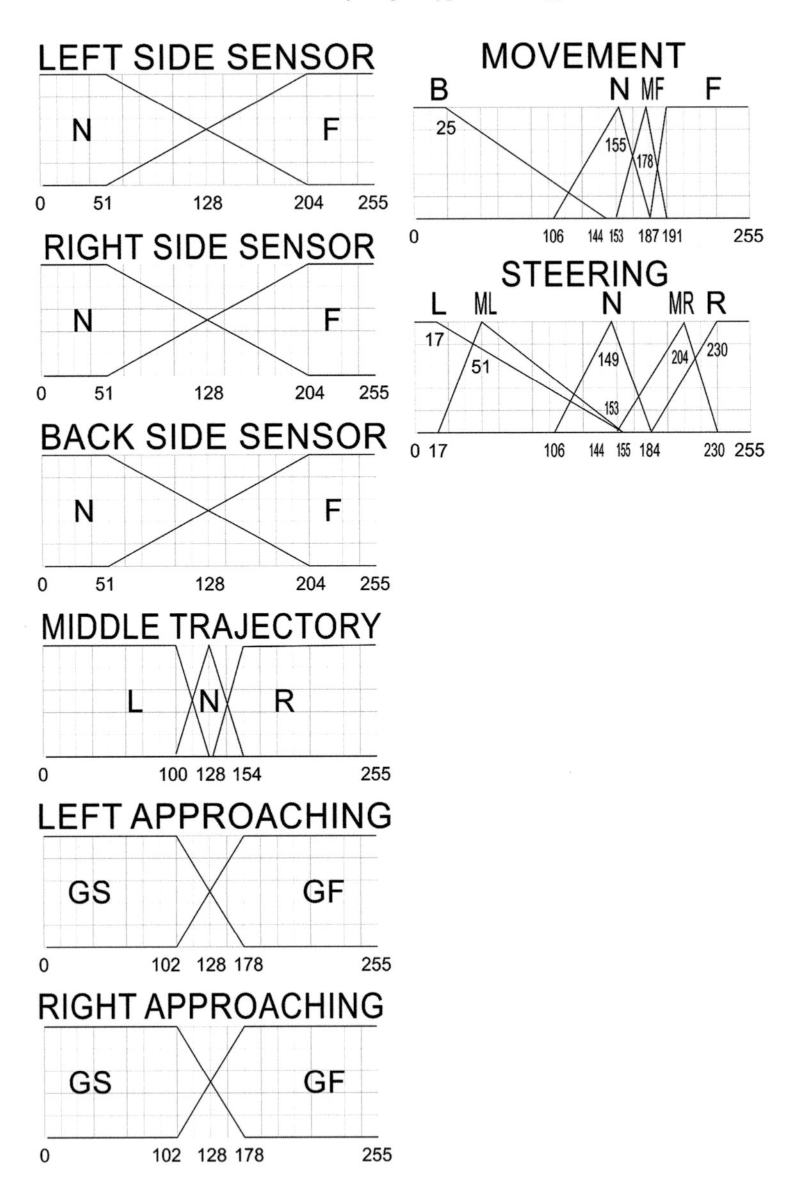

Fig. 4.31 Set distribution for all the variables of the wheelchair T1FLS

4. Now that all the variables, sets, and rules are defined, we proceed to calculate the crisp output based on the *firing strengths* f_i, the *T1 inferred set* the consequent T1FS *shape* using the *Centroid* defuzzification method approaching fast, we just use the GF set of each variable.

Finally, the T1FLS for the electric wheelchair is designed as you can see in Fig. 4.32, with the same program functional blocks, just as the servomotor example.

The resulting T1FLS is then connected in a control system as you can see in Fig. 4.33, where the magnitude converters are also used.

The purpose of this section is to show the user the ease of building T1FLS with the functional blocks provided in the program. The following section provides a guide for building T2FLS; the servomotor study case is retaken and from this a T2FLS will be created.

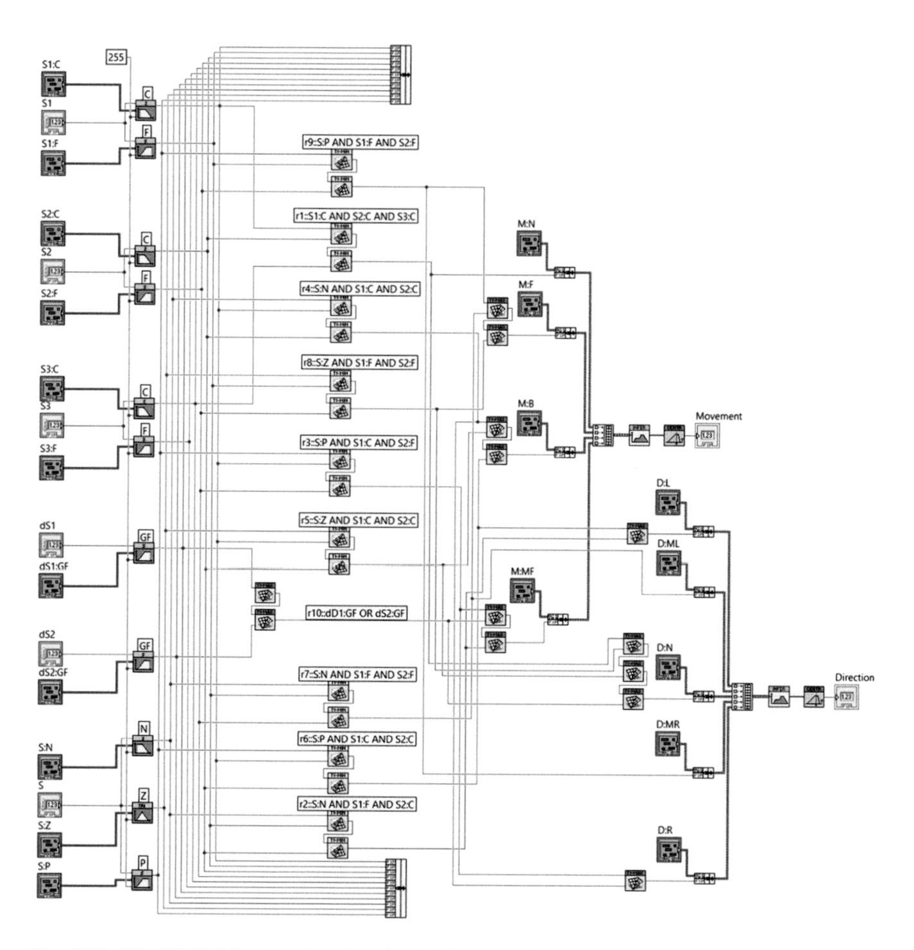

Fig. 4.32 The T1FLS for the electric wheelchair control

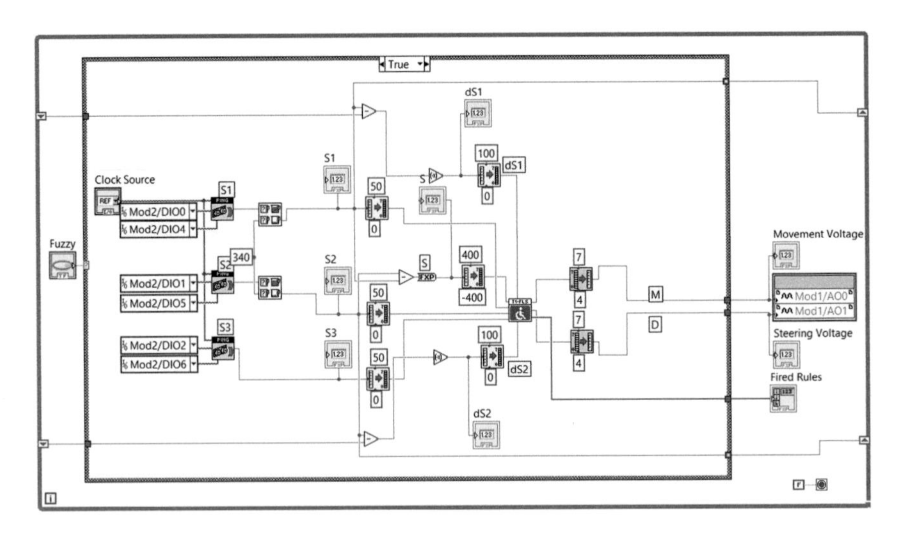

Fig. 4.33 The T1FL control system for the electric wheelchair

4.8 T2FLS Validation

DC servomotor

Implementation

As mentioned in the previous section, the servomotor study case is retaken and some additional parameters are included for the T2FLS implementation: the uncertainty widths (as stated in the Chap. 1) in order to obtain the FOU.

This FOU is what makes the difference between the T1 and T2FS. Its purpose is to absorb all the uncertainties due to the set and rule definitions.

All the figures and details about the target FPGA, servomotor details, connection diagrams, and inclusively the same rule set will be considered in this section. The only difference will be the FLS used.

See Fig. 4.26. You realize that each set has a parameter set for defining its MF. Now, for each parameter set you will have an uncertainty width set. This difference can be seen in Fig. 4.34, but it is not the only difference. Notice that each functional block dedicated to the FS is a T2FS functional block. This means that a T1 triangular MF can be replaced by a T2 triangular MF, so the other functional blocks. But if you change each T1FS for T2FS, you will need to add the uncertainty width sets, and each normalization value now is a cluster of two normalization values, one for each UMF and LMF.

Also, the conjunction and disjunction operations must be interchanged, e.g., all the T1 minimum intersection are replaced for their corresponding T2 minimum intersection functional blocks.

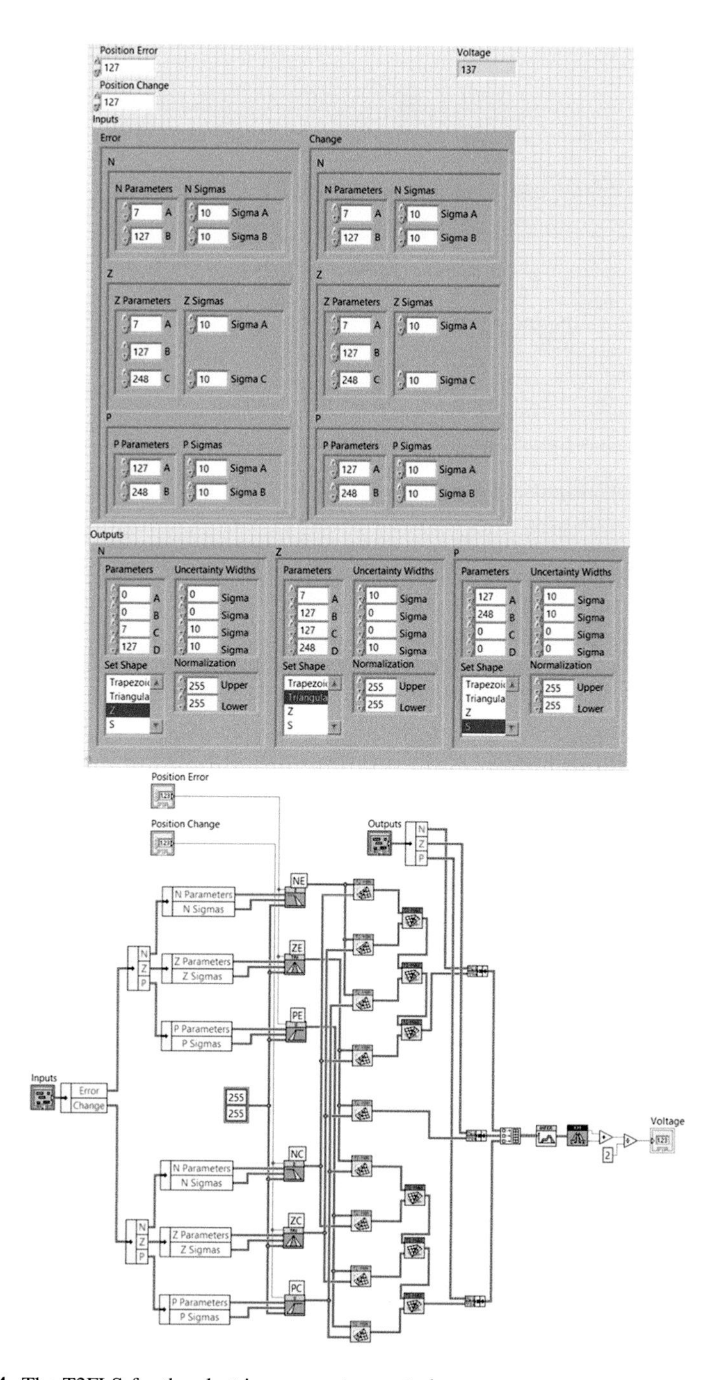

Fig. 4.34 The T2FLS for the electric servomotor control

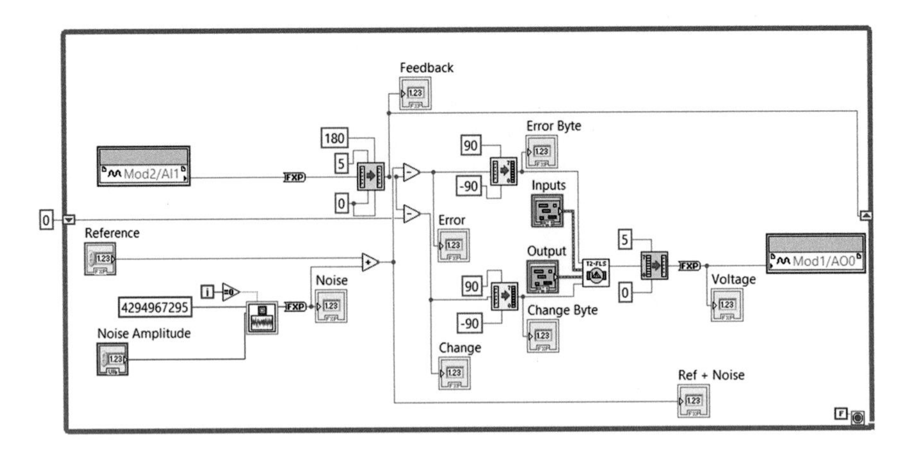

Fig. 4.35 The T2FL control system for the servomotor

The T1 inferred set is also replaced by a *T2 inferred set*. Also, the parameters that define the consequent T2FS must include the uncertainty width set and normalization as the fuzzification part needed it.

The Centroid fuzzification method is also replaced by the *Karnik–Mendel Algorithm* functional block. Due to the reason that its output delivers two Centroids, all you have to do is to average them. So, this will be the crisp output.

The T2FLS now can be included to the same control system (Fig. 4.27) as shown in Fig. 4.35.

Notice that both Figs. 4.27 and 4.35 have a white noise generator and a noise amplitude control. This is because the purpose of this section is to show the advantages of the T2FLS against the T1 and it can be discussed based on the results obtained.

4.9 Performance T1 FLS DC Servomotor

An additional VI was created and added to the project as a host VI called Graph, the Triangular Response.vi. This VI synchronizes its operation with the target FPGA, i.e. the cRIO-9014 and the target VI called T1 DC Servo Control.vi. This VI modifies the value of the reference in the target VI and describes a triangular reference, which starts from π, decreases, and when the reference is 0, then starts increasing until π is reached again. This way, the proposed T1FLS (Fig. 4.26) can be plotted in which presents a great performance (Fig. 4.36), so the response is very close to the reference. Notice that the noise present is inherent to the data acquisition and the control noise amplitude is set to zero (no additional noise is added). This control is reserved for the next section.

The set distribution was calculated by trial and error.

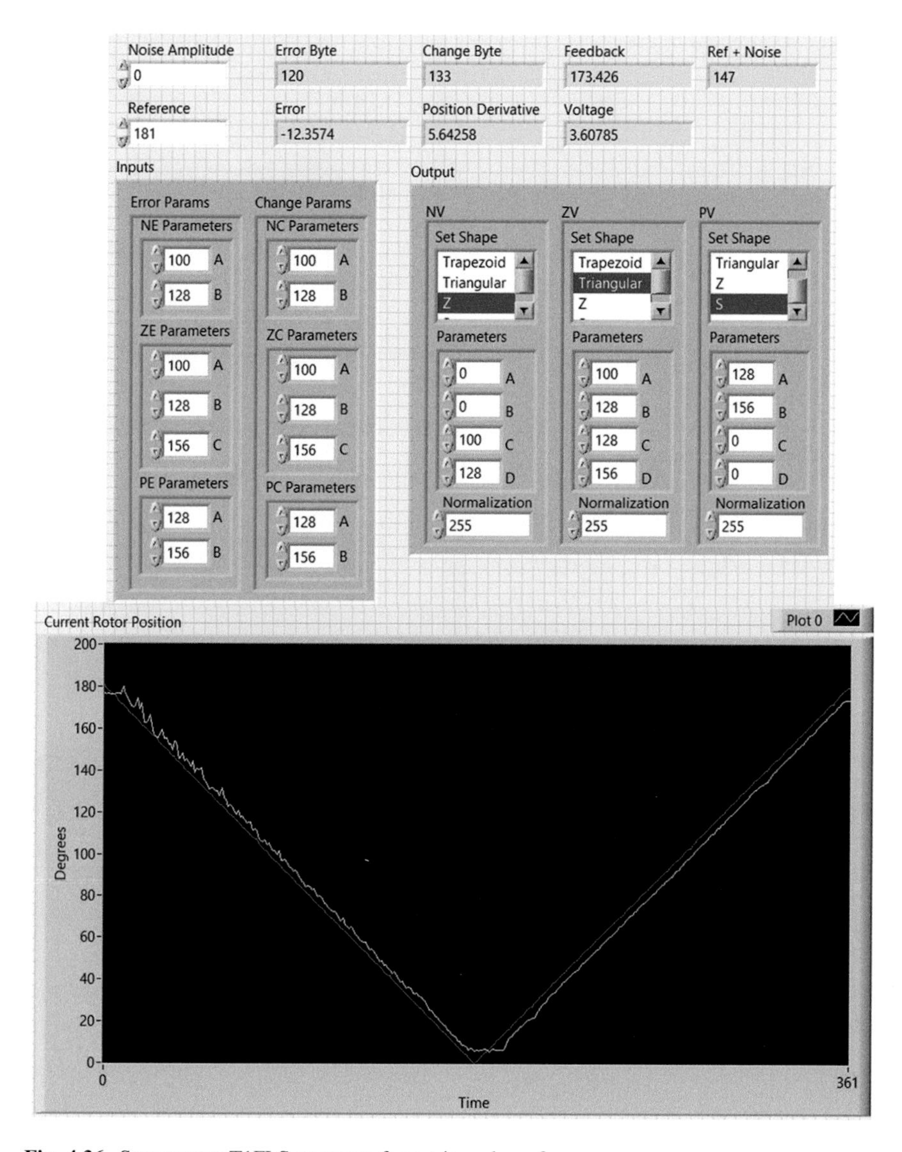

Fig. 4.36 Servomotor T1FLS response for a triangular reference

4.9.1 Electric Wheelchair

The electric wheelchair was tested in a square maze where it should enter by one side and exit by the other side without colliding with the walls. Its dimensions are shown in Fig. 4.37. Its response can also be found in Fig. 4.38.

Their set distribution was also calculated by trial and error. So, its performance can be improved if an optimization algorithm is used.

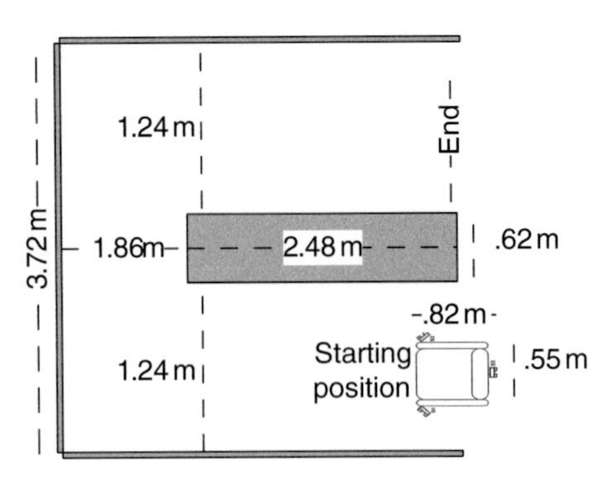

Fig. 4.37 Dimentions of the square maze

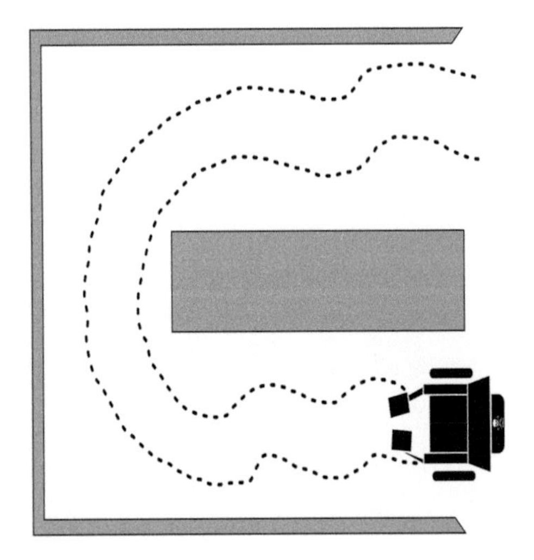

Fig. 4.38 Electric wheelchair T1FLS response in a square maze

The following section provides the T2FLS performance verification, where only the servomotor case is considered. After this section, the user will understand the advantages of using T2FLS instead of T1.

4.10 T1FLS Versus T2FLS

4.10.1 Noise Response

For this section, only the servomotor is considered due to its simplicity. Only the control systems considered are the ones shown in Figs. 4.27 and 4.35. For this case, the reference is also modified with the same triangular shape as Fig. 4.36, but now some white noise is aggregated to it in order to introduce some uncertainty. The noise amplitude control available in both T1 DC Servo Control.vi and T2 DC Servo Control with KM.vi have the objective of introduce this uncertainty in the system.

Additionally, the T2 DC Servo Control with NT.vi is provided for comparison purposes.

Now, let us see how T1FLS and T2FLS behave in the presence of noise. Finally, you will see the importance of the FOU whenever you change the uncertainty widths for each parameter.

Notice, in Fig. 4.39, how the noise amplitude control in T1 DC Servo Control.vi, aggregates noise to the reference. The T1FLS should deal with the uncertainties due

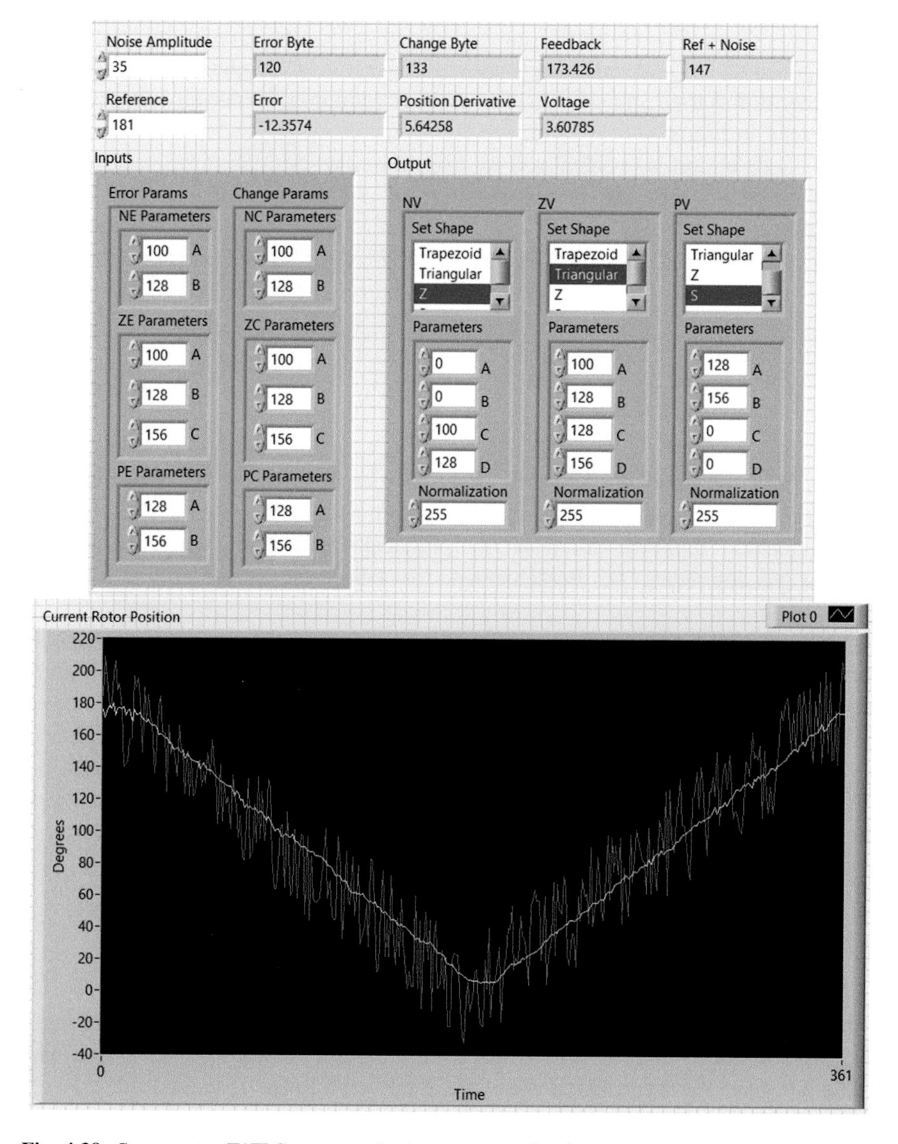

Fig. 4.39 Servomotor T1FLS response in the presence of noise

to the linguistic concepts but not necessarily with those related to the set definition and then if, the crisp inputs present uncertainties. Observe that some regions are crispy and some delay is introduced to the current position.

The T2FLS, in this case should deal also with those uncertainties in the set definition. Then, when a crisp input presents uncertainties, i.e., noise, then the T1FLS should take bad decisions. So, the T2FLS is capable of dealing with this and take better decisions.

In figure 4.40, the T2FLS clears the noisy reference with the FOU and takes better decisions improving its response and achieving better position curves. This T2FLS is performed using the NT method as defuzzification. Due to the reason that KM Algorithm is too expensive in timing response, the NT method was used. If the KM Algorithm is used, considering the same set definition, its response is worse, as can be seen in Fig. 4.41.

The FS definition was calculated by trial and error. An optimization algorithm can be used to adjust the best parameter set and as a consequence, the best performance.

It is recommended to start adjusting the T1FLS set parameters and from this optimal configuration the FOU widths can be tuned in the presence of noise.

4.10.2 Response Time

During every iteration, the KM Algorithm searches a centroid; this iterative KM version is very slow, performing an inference each 51 ms, i.e., *19.62* FLIPS, the full KM Algorithm implementation is not practical for real-time applications. Its reduced version must be implemented for being feasible for this kind of applications.

Unlike the KM Algorithm, the NT method lasts just 1.82 ms, so it can respond better in this application and can be compared to the T1FLS response. The T2FLS implemented with the NT Method can perform 549.45 FLIPS.

By its complexity, the T2FLS are slower than the T1 if the full KM Algorithm is used, since more data is processed per inference. But, with the NT Method the T2FLS can compete against the T1FLS, which lasts 1.41 ms per inference, just slightly faster.

4.10.3 Resource Utilization

Table 4.4 presents information related with the FPGA resource utilization for each functional block, Table 4.5 shows additional utilities included in the toolkit and Table 4.6 gives the information about the slices used in the functional blocks.

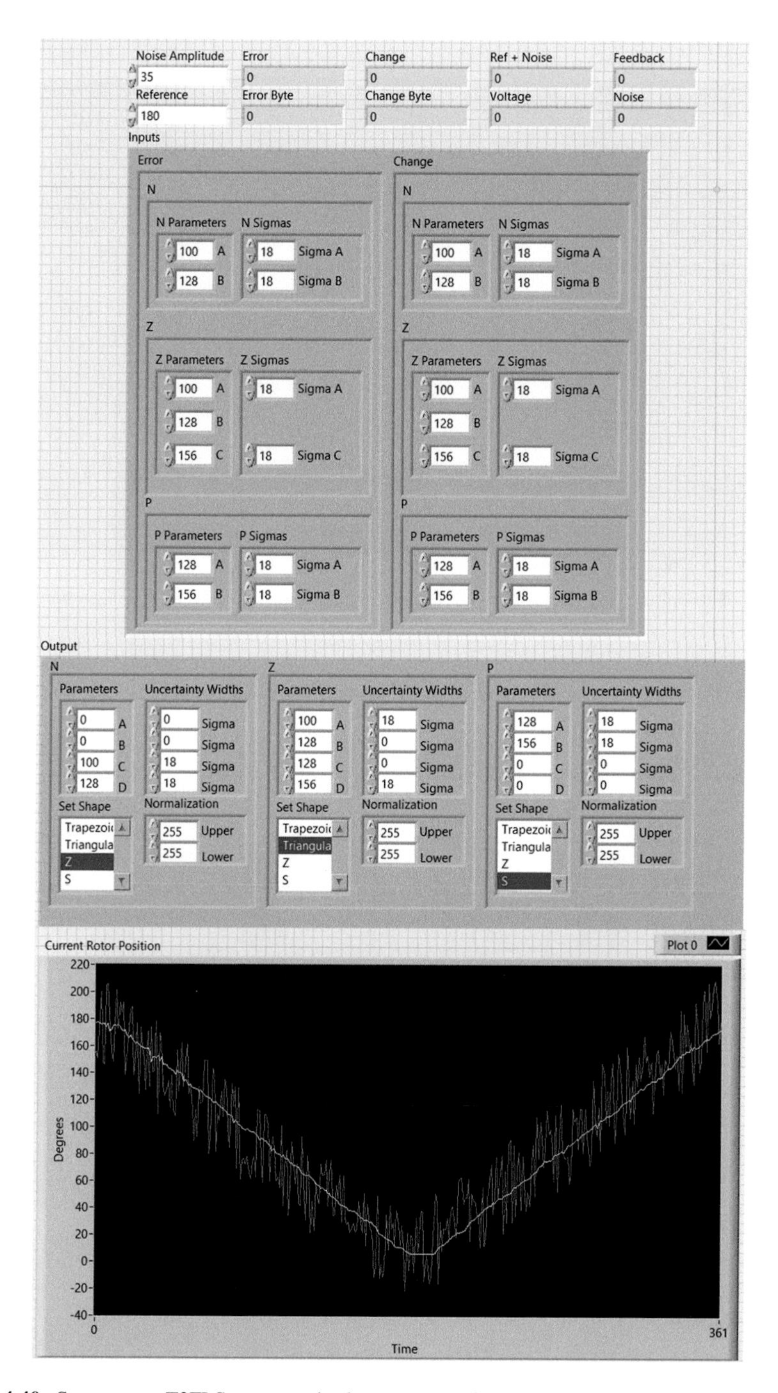

Fig. 4.40 Servomotor T2FLS response in the presence of noise using the NT method

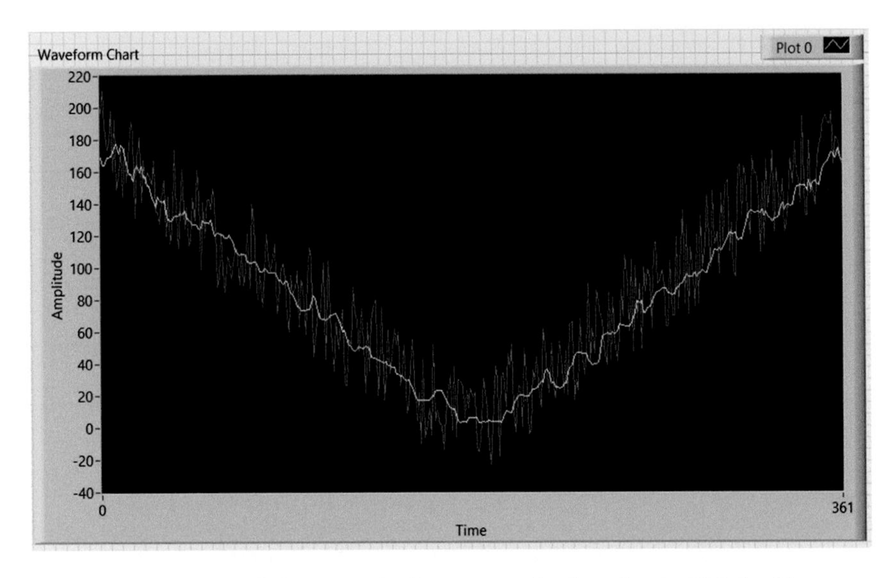

Fig. 4.41 Servomotor T1FLS response in the presence of noise using the KM algorithm

Table 4.4 FPGA resource usage for several T1FLS and T2FLS design blocks and the proposed applications

Functional block	FLS				T2FLS			
FPGA resources	Logo	Total slices	Slice registers	Slice LUTs	Logo	Total slices	Slice registers	Slice LUTs
S-shape		659	734	813		846	970	1056
Trapezoidal		891	997	1117		1280	1437	1623
Triangular		830	924	1037		1175	1287	1490
Z-shape		661	734	814		840	963	1050
Support		729	729	1036		725	726	1034
Minimum intersection		426	513	522		525	596	656
Product intersection		426	522	506		530	612	627
Bounded intersection		512	567	650		572	669	720
Drastic intersection		518	552	641		570	650	723

(continued)

Table 4.4 (continued)

Functional block	FLS				T2FLS			
FPGA resources	Logo	Total slices	Slice registers	Slice LUTs	Logo	Total slices	Slice registers	Slice LUTs
Maximum union	T1-MAX	471	515	605	T2-MAX	523	594	656
Algebraic sum union	T1-SUM	485	543	605	T2-SUM	522	648	613
Bounded union	T1-BOU	458	540	546	T2-BOU	544	629	676
Drastic union	T1-DRA	462	551	564	T2-DRA	568	650	718
Inferred set	INFER	2540	4126	2728	INFER	3500	5728	3858
Centroid	CENTR	646	818	830	–	–	–	–
Non-iterative Karnik–Mendel algorithm	–	–	–	–	N-KM	1461	1454	2185
Iterative Karnik–Mendel algorithm	–	–	–	–	I-KM	2415	2087	3759
Enhanced Karnik–Mendel algorithm	–	–	–	–	EKM	2593	2828	3965
Nie–Tan method	–	–	–	–	NT	915	959	1305
TSK	T1-TSK	2997	4344	3670	T2-TSK	8391	10075	11705

Table 4.5 Utilities

Functional block	Logo	Total slices	Slice registers	Slice LUTs
Magnitude to byte		1033	1219	1290
Byte to magnitude		910	1094	1197
Magnitude to magnitude		1184	1518	1537
Parallax PING))) sensor decoder	PING	789	868	1053
Polynomial generator	POLY-2	671	881	710

Table 4.6 Applications

Functional block	Logo	Total slices	Slice registers	Slice LUTs
T1 DC servomotor	NA	3849	4299	5011
T1 DC servo control	NA	7234	7482	10,442
T2 DC servomotor with KM	NA	7761	8101	10,710
T2 DC servomotor with NT	NA	7045	7564	9553
T2 DC servo control with KM	NA	10,934	11,348	15,816
T2 DC servo control with NT	NA	10,381	10,810	14,897
T1 wheelchair	NA	7017	7741	9162
T1 wheelchair control	NA	9794	9562	14,888

4.11 Included Examples

The FPGA fuzzy logic program includes the application examples for both T1FLS and T2FLS used to explain this program; they are located in examples\chapter5 directory. These examples let you implement and interact with several parameters in order to compare both T1 and T2 performances.

You will have to copy the entire project directory to your PC and open the project. For this purpose, you will have to use the same cRIO device and connect the C-series modules in the same slots as defined in the project and configure your device via MAX (Measurement Automation Explorer).

If you want to use another device, you will have to create a new project and install the C-series module or the pertinent tools for acquiring and provide analog data for the applications. Finally, you will have to copy and paste the entire user Vis in the directory.

4.11.1 Case Study: Experimental CNC Micromachine Controlled by Fuzzy Type 2

Since micromachines have been changing according to new industrial requirements, a swift technological evolution is required. For instance, a lot of studies about micromilling processes have shown that new manufacturing demands are increasing drastically and the real-time controllers play new roles to fullfill those necessities. Intelligent control systems can deal with manufacturing problems as long as adequate hardware or software platform is selected for implementing the controller's algorithm. Hence, Field Programmable Gate Arrays (FPGAs) [1, 2] are key elements for deploying intelligent controllers because they can make very fast operations and can include bidirectional data buses. One of the first controllers implemented in FPGAs was a linear controller but it could not cope with uncertainties that were presented inside the system. In micromilling [3], conventional controllers were applied and they showed good results under linear operation points [4], yet in some industrial systems do not have ideal conditions in complete operation range, so the linear controllers cannot get the manufacturing requirements in each operation range. Hence, the micromachine systems could not achieve high precision in position loops. Noise and uncertainties in manufacturing systems are quite serious problems in the position, torque, or speed loops. They can show up when the environment of manufacturing process is corrupted by electronic and digital signals, which generate noise in all the frequencies, and the noise isolation in the environment is not applied because of its expensiveness. On the other hand, the manufacturing machines require high precision for producing high-quality products [5]. Fortunately, advanced control algorithms can improve the performance of the manufacturing machine when noise and uncertainties are in the process. The manufacturing machines have to achieve high standards. For instance, high precision in dimensions (between 1 and 999 μm) have to be reached in a correct manner even if the noise signal occurs. Additionally, the CNC machines are designed for accomplishing different goals such as speed, precision, metal cutting volume per unit time, and so on. The controller's law and the real-time platform have to be selected according to the mechanical design of the micromachine. If low-cost position sensors without noise isolation are used in the CNC micromachine, the input signal of the controller is degraded and the precision of the complete performance of the micromachine is diminished unless the controller implemented can tolerate noisy conditions. As a result, PID conventional controllers are not a good option under those conditions for they are not able to deal with uncertainties or noisy conditions. Usually, conventional manufacturing involves developing specific work-pieces using regular machine tools, and they use conventional controllers. However, nonconventional manufacturing systems such as ultraprecision machining methods, which get high accuracy in dimension standards between 0.05 and 0.005 μm [6], need advanced controllers such as intelligent controllers. This kind of application requires reducing the indexes of performances by advanced control techniques. Moreover, high-speed machines require decreasing the machining process time and increasing the high-quality production [6]. In the case of micromachining, the machine tools are very small (around 1 μm overall dimensions) and the work-pieces are available with tolerances of 10 and 0.01 μm, using high-speed spindles, usually above 10,000 rpm. The general problem in those machines is to improve the contour tracking that is linked with the position loop [7]. Nevertheless, the industrial goals imposed by miniaturization of machines such as speed control loops, position control loops and torque control loops created a very complex control system, which requires normally high efforts of the controller. The micromachining procedure focus on mechanical microcutting techniques in which the material is removed by cutting the edges; thus, the fuzzy logic controllers proposed are designed for accomplishing this kind of mechanical tasks. The aim of this paper is to generate a fuzzy logic type 2 controller in real-time FPGA, which copes with noisy signals and uncertainty conditions presented in manufacturing process. Generally, those conditions are presented in industrial environment and the conventional controllers such as PID controllers are not able to cope with those circumstances. The paper also shows the advantages of using real-time FPGA in manufacturing CNC machines. When the proposed fuzzy logic type 2 controller is deployed in the CNC micromachine under different cutting trajectories, the real-time FPGA platform runs the algorithm in deterministic time that increases the controller's performance. Furthermore, the paper presents a complete analysis about the indexes of control performance for linear PID, fuzzy logic type 1 and fuzzy logic type 2 in order to select the suitable controller according to the manufacturing requirements.

4.11.2 Micromachines and Fuzzy Logic

The micromachines tendency is to develop manufacturing systems with the ability to make small pieces (below 4 mm in diameter) in a single setup at low cost [6]. The method for machining small pieces is based on sliding head-stock machines because they are the most efficient manner for producing large volumes. This name was coined from the fact that the sliding head-stock machine was equipped with a main head function, which uses a guide bush and the material can slide in the Z-direction in a system of three axes. Hence, high stability condition has to be achieved. Most sliding head machines are CNC-controlled and they can deal with a maximum of 32 mm bar diameter [6]. The advantage of using fuzzy logic is to emulate the human knowledge expressed in linguistic rules; so it is possible to incorporate fuzzy logic systems when some CNC machines have uncertainties and/or noise. Zadeh in [8, 9] concluded that fuzzy logic sets are adequate for modeling the human perception including uncertainties. The Type-2 Fuzzy Set (T2FS) has a fuzzy domain; this allows improving the ability to cope with uncertainties. Depending on the manufacturing trajectories and noisy signals, the controller selected shows different responses under those conditions. A measured signal from sensors in a CNC micromachine is normally composed of two components, one is the meaningful signal, the other is the undesired noise signal, and the measured signal is then given as the sum of those signals. The measured signal is one of the inputs to the controller, so in this paper noisy signals are added in the input signal of the controller. Thus, the controller has to cope with those conditions to get good manufacturing results. Fuzzy logic type 2 is well known for tackling noisy conditions [10, 11] and real-time FPGA have been used for deploying advanced controllers. The complexity of the fuzzy logic type 2 is decreased by type-reduction type 2 which was proposed by Karnik and Mendel [12] that is an iterative algorithm for calculating the approximated centroid of an Interval T2FS (IT2FS). Because of its simplicity, type-reduction algorithm is applied in a large number of industrial application and fuzzy logic controllers based on type 2 have solved different manufacturing problems [12]. Mendel and John proposed an easier way to represent the T2FS in a three-dimensional function, comprised by the FOU, which is based on the primary and secondary membership functions that represent the probability distribution as it was presented in [11, 13, 14]. Furthermore, Wu and Mendel in [15] proposed a formula to map an IT2FLS using the uncertainty bounds in an easy manner. Some additional modifications for calculating the output of the fuzzy controller type 2 claimed to decrease the computational complexity of the IT2FLS [11, 16–19], and they are used for implementing IT2FLS real-time platforms. The NT Method [15, 20] is applied to FOU sections, if the FOU does not present inflections; in fact, the FOU inflection must be specially computed by radial projections from the LMF to the UMF and vice versa [21–23].

4.11.3 Reconfigurable Micromachine Tools

The Reconfigurable Micro Machine Tool (RmMT) has mechanical, electrical and digital control components. The RmMT is designed to achieve three different modes and it is configured by changing the basic elements and the cutting tool. The RmMT configurations are: Lathe, Milling Machine, and Drill [1]. The elements of the RmMT are made of aluminum 1060 and were designed to be adjusted easily and the mechanical parts do not let deformations in the mechanical structure. The volume measurements for vertical configuration in the Milling Machine or Drill are 216 mm × 280 mm × 220 mm, and horizontal dimensions are 216 mm × 280 mm × 110 mm. The measurements for the Lathe are 216 mm × 280 mm × 110 mm. The following list shows the elements needed to assemble the RmMT Kit (see Fig. 4.42).

1.	Principal base	11.	Linear actuator
2.	Support for a vertical reconfiguration	12.	Portable actuator
3.	Reconfigurable base	13.	Clamp cylinder
4.	Tower support jaw/spindle	14.	Cutting tool (milling machine or drill)
5.	Angle (support of axis *Y*)	15.	*X*-axis motion actuator
6.	Base for support of axes	16.	*Y*-axis motion actuator
7.	Spindle screw clamp support base	17.	*Z*-axis motion actuator
8.	Spindle screw clamp	18.	Burin
9.	Cutting tool support	19.	Spindle
10.	Clamp jaw	20.	Support burin

Linear actuators [24], based on direct current motors, take over the displacement of the micromachine tool in order to reach the reference positions. The main elements of the control system for the RmMT are position sensors (Encoder), limit sensors (Hall Effect), and hardware real-time platform (FPGA) [2]. In this case, an

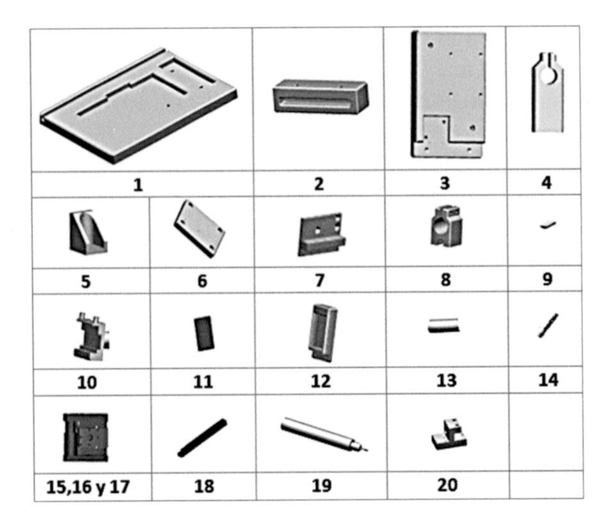

Fig. 4.42 Elements that involve the assembly of RmMT

accurate tracking performance is essential part in ultra-precision devices [7]. For this reason, a high performance controller must be included as a core part in the micromachines. With the purposes of this study, the machine was adapted in the lathe configuration. The lathe configuration is the simplest one in the RmMT system, so it uses a minimum number of elements for assembling the machine tool and it always runs in the horizontal axis. This configuration allows including two axes of motion [1] (see Fig. 4.43).

In this configuration, the RmMT can manufacture work-pieces by angular displacements, while the actuators move the cutting tool in a controlled manner against the surface of the work-piece. The work-piece is cut according to the numerical control specifications [1]. The RmMT has linear motion systems that have direct

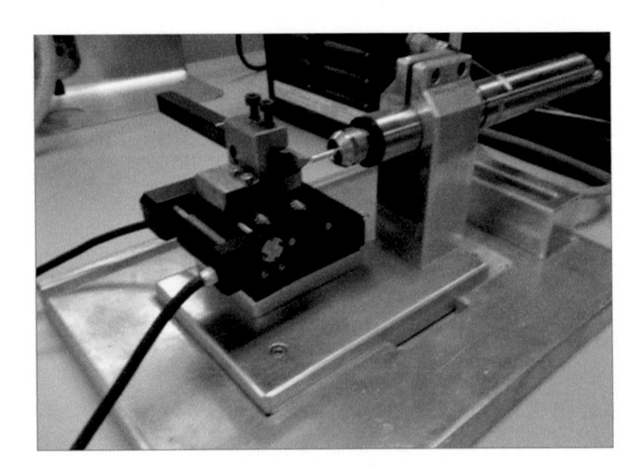

Fig. 4.43 RmMT system in lathe configuration

Table 4.7 Technical specifications

Parameter	Value
Motor type	DC motor
Travel range *X*, *Z* axes	(15 × 15) mm
Design resolution *X*, *Z* axes	0.0085 μm
Max. velocity *X*, *Z* axes	2.2 mm/seg
Nominal motor power *X*, *Z* axes	2.0 W
Motor voltage range *X*, *Z* axes	0 to ±12 V
Spindle	Pneumatic motor
Max. velocity	30,000 rpm
Max. motor power	125 W
Collet chuck	Ø 0.5–6.0 mm
Power consumption	92 W

current (DC) motors. The ultra-precise microactuators give high translation stages that provide linear movement from 0 to 15 mm in a compact package. Table 4.7 illustrates the specifications of the RmMT.

4.11.4 Motion Control

Control of motion is required in several industrial applications, which involve precise position and speed control loops [23]. As result, there are motion controllers that run intelligent control techniques to accomplish complex position profiles [25]. Indeed, this CNC machine tool is designed to get high precision in position using real-time FPGA because FPGA can send control signals at high speed. Some research papers have been focused on the implementation of new motion controllers in CNC machine tools. For instance, Professor Koren [26] evaluates servo-controllers and presents different variables of machining process when the status of the machine tool is considered into the design process. Yoram classified servo-controllers as it is shown below but the paper does not study fuzzy controller's that are excellent control systems. Besides, the fuzzy logic controllers are based on linguistic rules that can be understood by human operators [27].

- Feedback controllers (P, PI, PID, state-feedback)
- Feed-forward controllers (ZPECT, IKF)
- Cross-coupling controllers (CCC)

Position controllers of servo system are key elements, which determine the precision and efficiency of the manufacturing process; indeed, it is recommended to conduct mechanical simulations and to design a controller after an experimental

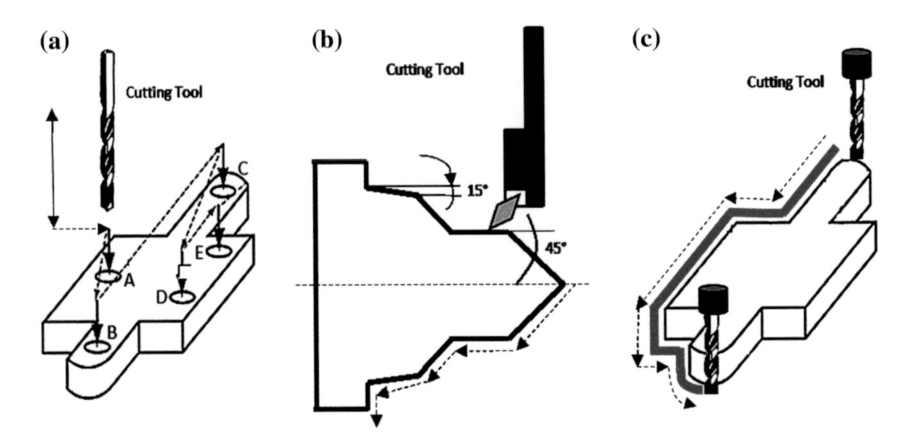

Fig. 4.44 Control strategies of CNC machine tools. **a** Point to point control. **b** Tracking control. **c** Contour control

design could be concluded in manufacturing process [26]. Conventional machine tools started the evolution of micromachine tools, which are presented in [11]. Generally, the control strategies in CNC machine tool are: (a) point-to-point control, (b) tracking control, and (c) contour control (see Fig. 4.44) in which the command signal is the desired position of each axis. For instance, one of the simplest conventional linear controllers is the proportional controller in which the output signal is proportional to the error input signal but conventional linear proportional controllers are not good alternatives for CNC machines under uncertainties and noisy signals. In a micromachine tool, a particular controller handles each axis of movement (see Fig. 4.44). Moreover, it is also required to ensure reasonable precision and high robustness against disturbances during the close loop operation (see Fig. 4.45).

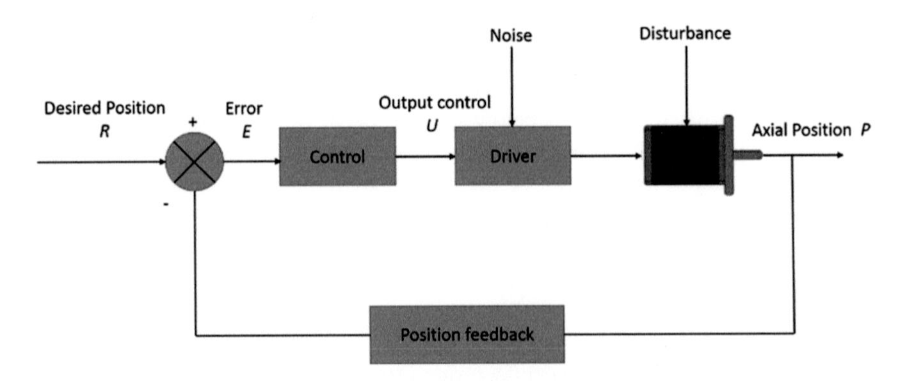

Fig. 4.45 Position control closed loop for machine tools

4.11.5 Control Design on Real-Time FPGA

There are some implementation of fuzzy logic controllers that are based on conventional proportional-integral (PI) controllers but most of them are focused on fuzzy type 1 [28]; thus, a complete control design for manufacturing process has to be done for fuzzy logic type 2. In general, position control of manufacturing machines used as the main actuators in the position loop DC motors and real-time systems. The real-time FPGA allows improving the tracking of the reference position signals. The changes in the torque load conditions generated by the cutting process (mechanical disturbances for the electric motor) have to be supported by the controller. This paper proposed a position controller for CNC micromachine that is based on fuzzy type 2; the controller is able to deal with noisy conditions and uncertainties [29]. On the other hand, the real-time FPGA hardware platform selected guarantees that control loops can be completed by a given limit period of time under deterministic conditions and the control algorithm was done in a graphical programming (LabVIEW FPGA); thus, reviewing the controller code is easy [30].

The first controller designed in this paper for controlling position on real-time FPGA was a discrete PID in a parallel topology that is implemented by its digital representation [31]. The discrete definition of the PID controller could be derived by the continues mathematical expression shown in Eq. 4.1.

$$\mathrm{PID}(s) = e(s)\left(K_p + \frac{K_i}{s} + K_d s\right) \tag{4.1}$$

where K_p, K_i and K_d are constant PID values that have to be tuned for accomplished high control standards [31–33]. A discrete mathematical transformation was done by Tustin method [31] to get the digital expression illustrated in Eq. 4.2.

$$\begin{aligned} K_B &= \frac{2K_iT_s^2 - 8K_d}{2T_s} \\ K_C &= \frac{K_iT_s^2 - 2K_pT_s + 4K_d}{2T_s} \\ \mathrm{PID}(z) &= e(z)\left(\frac{K_Az^2 + K_Bz + K_C}{z^2 - 1}\right) \end{aligned} \tag{4.2}$$

Equation 4.3 shows the recurrence expression for the PID controller.

$$\mathrm{PID}(k) = K_Ae(k) + K_Be(k-1) + K_Ce(k-2) + \mathrm{PID}(k-4) \tag{4.3}$$

where $K(A, B, C)$ are the constant values of the PID controller at the current sample time at T_s interval and the PID tuned values for CNC micromachine are

$$K_p = 32.2305, \quad K_d = 0.00499939, \quad \text{and} \quad K_i = 0.0289997 \tag{4.4}$$

The fuzzy logic controller type 1 is a particular case of a fuzzy logic controller type 2 [27] when the uncertainties disappeared, so the same design process is applied to fuzzy type 1 and type 2. In fact, the fuzzy logic controller type 1 is reached by removing the FOU. The main blocks in fuzzy type 2 are shown in Fig. 4.46a in which the selected inputs are the error and change in the position and

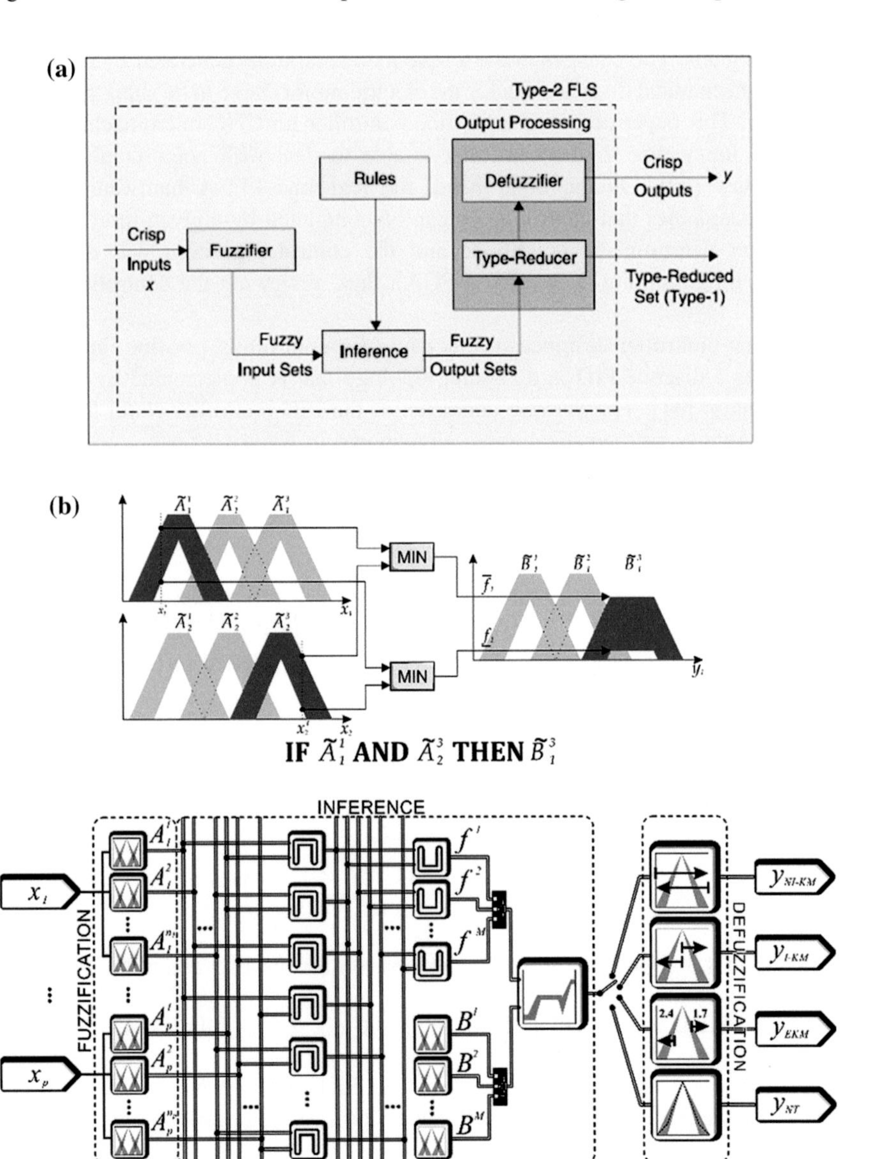

Fig. 4.46 **a** Fuzzy logic type-2 and **b** fuzzy inference

Fig. 4.46b illustrates the fuzzy inference. Practically, fuzzy logic type 1 has the same control blocks but it does not include the type reduce set block which is located in the output processing diagram (see Fig. 4.46a).

The implementation of the fuzzy logic controller in terms of type 2 fuzzy sets, could be done by two inputs, the error e(t) that is the difference between the command signal and the feedback signal from the position sensors, as well as the position change $\Delta p(t)$. The complete description of the fuzzy logic type 2 is presented in Fig. 4.47; if the uncertainty is eliminated, a fuzzy logic controller type 1 is gotten. The linguistic rules implemented in the fuzzy type 1 and type 2 are illustrated in Fig. 4.48; those rules are obtained according to the trajectory described in each case of the universe of discourse presented by the position control system (see Fig. 4.48). Not only can nine rules cover the complete linguistic universe of discourse for a position controller, but also the number of linguistic rules allows to keep a simpler control algorithm that is easy to implement in a real-time FPGA platform, so the fuzzy logic controller is based on simple and compact linguistic expressions. In addition, the FPGA runs with basic mathematical operations the fuzzy logic algorithm and the position controller's performance is improved.

For deploying the controllers, the cRIO-9014 [34] were used, which comprises a real-time microcontroller, a FPGA backplane chassis, and two C-Series modules NI-9505 [29, 35] (see Fig. 4.49). The FPGA can acquire the necessary information from encoders, can process the controller algorithm and can generate the output voltage signals by pulse width modulation (PWM). cRIO has a finite number of resources such as multiplications, memories locations, and so on. Thus, the resources and processing time used in fuzzification, which is the defuzzifier (see Fig. 4.46a), are key elements that have to be analysed in order to improve the performance of the fuzzy logic controller. Some research works show results about decreasing the complexity of defuzzification step but a specific study is obligatory for using the cRIO-FPGA in a correct manner. The Karnik–Mendel (KM) algorithm

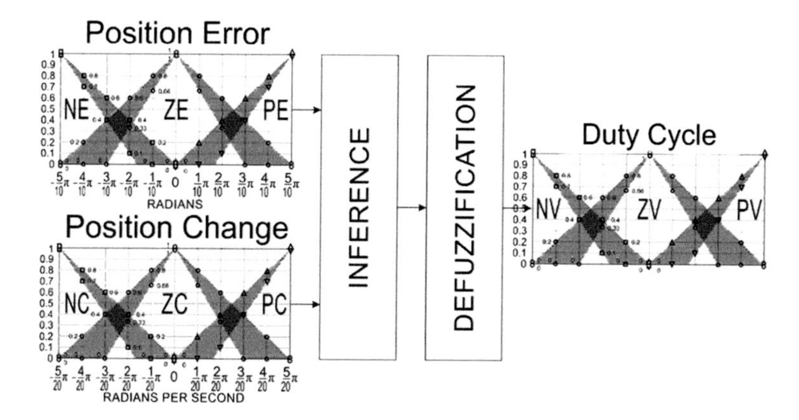

Fig. 4.47 Fuzzy type 2 inference controller

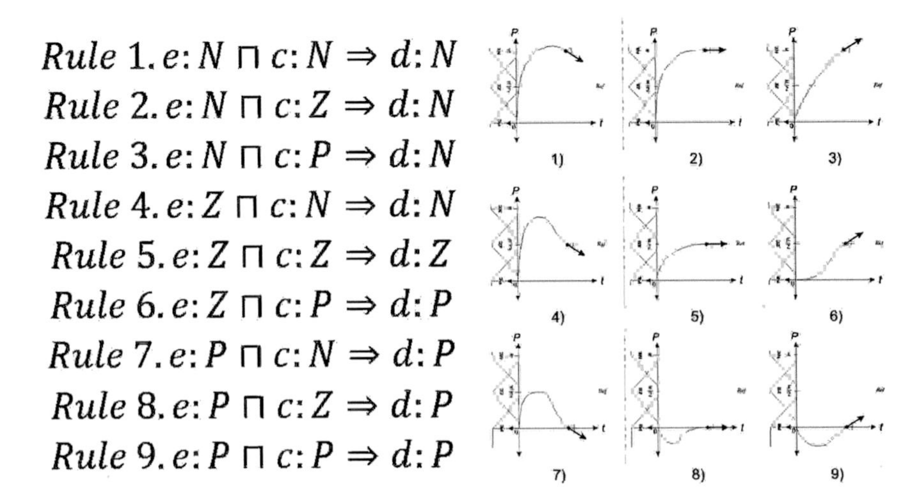

Fig. 4.48 Rules fuzzy type 2 inference controller

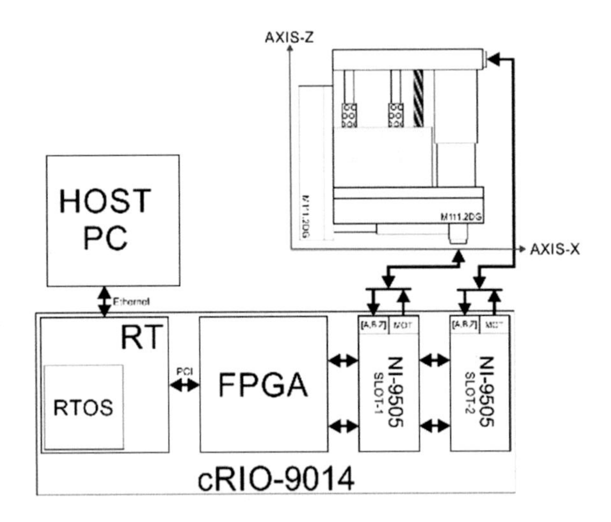

Fig. 4.49 c-RIO topology for the CNC position controller

[36] and the Nie–Tan (NT) method [19] were designed, implemented and compared according to LabVIEW FPGA basic block diagrams.

After the fuzzification algorithms were deployed in the cRIO, the FPGA resources and time conditions were calculated for selecting the fuzzification structure. As a result, the enhanced Karnik–Mendel algorithms (EKM) [18] is the most expensive method in computational terms and NT method is the fastest method which requires the minimum number of resources because it uses only one division and one multiplication to calculate the final centroid. Since this method does not search for several embedded sets, the final centroid is a single value. The EKM and the Iterative KM [37] require three units of centroids while the non-iterative version only requires two. In addition, the final time and the number of resources are presented in terms of Tick Count Blocks; the resources used are calculated by the Build Specifications module in LabVIEW FPGA [30]. Although the NT is not based on the KM algorithm, it uses an iterative version of KM algorithm in order to find the initial switching point, so the algorithm runs very fast. The NT is an algorithm that runs fast, it does not include complex mathematical blocks for calculating the fuzzification step, and it calculates the average calculation within the LMF and UMF, so the fuzzy description is acceptable for micromachines tools. As it is seen in Tables 4.8, 4.9 and 4.10, the NT is the simplest and the fastest defuzzification method for IT2FLS based on real-time FPGA; as result, the NT method was implemented in this paper [19].

Table 4.8 Number of iterations (N)

Element/method	Non-iterative KM [36]	Iterative KM [37]	EKM [18]	NT [19]
ES calculation	N	N	N/A	N
ES centroid calculation	$N + 3$	$N + 3$	$N + 1$	$N + 3$
Total iteration count	$2N2 + 4N$	$[N/2, 2N2 + 4N]$	$[1, 0.1764N2 + 2.1N]$	$2N + 3$

Table 4.9 Number of hardware elements used

Structure/method	Non-iterative KM [36]	Iterative KM [37]	EKM [18]	NT [19]
Multiplications	1	1	6	1
Divisions	1	2	5	1
Sums/subtracts	3	4	17	5
Centroid calculation units	2	3	3	1
Comparator/multiplexers (MUX)	5	9	9	1

Table 4.10 Timing performance and resources used

Resource/methods	Non-iterative KM [36]	Iterative KM [37]	EKM [18]	NT [19]
Latency (hardware) in milliseconds	49.48	0.8875	0.1756	0.27
Latency (software) in milliseconds	84.535	1.87	1.43	1.32
Slices	1461	2415	2593	915
Registers	1454	2087	2828	959
LUT	2185	3759	3965	1305

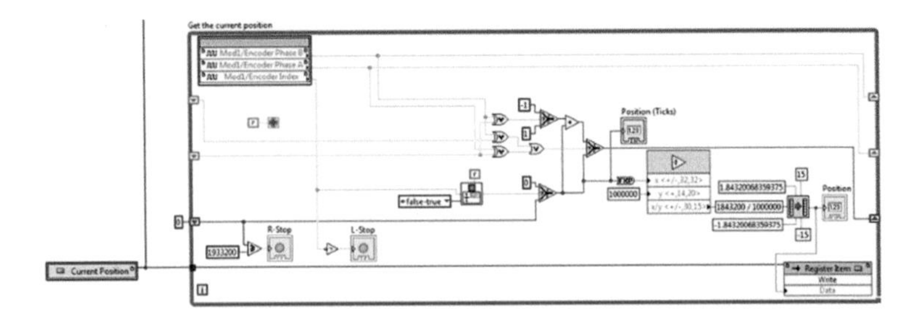

Fig. 4.50 LabVIEW program for reading the quadrature encoder by LabVIEW FPG

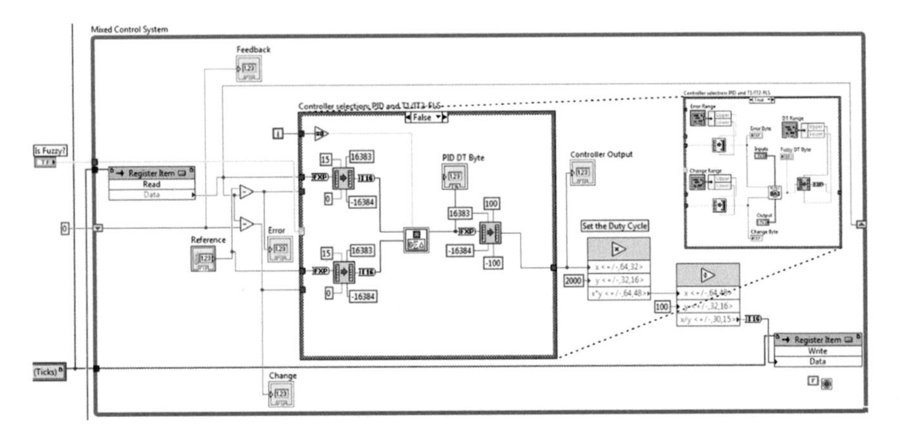

Fig. 4.51 LabVIEW program for controlling the position (PID or T1FLS/T2FLS)

The FPGA is the core element in the real-time system and the following tasks were programmed on it using LabVIEW real time.

1-To read quadrature encoder values for detecting the real position (see Fig. 4.50).

To run the controller's algorithm see Fig. 4.51.

To generate the output voltages signals, PWM voltage signals that drive DC motors (see Fig. 4.52).

Figures 4.53 and 4.54 illustrate the main block diagram for fuzzy type 1 and 2 and this block diagram is connected with the function panel for running the controllers under different reference signals and noisy conditions.

4.11.6 Experimental Results

This section presents how the position control loop is affected by noisy signals under different position commands. When a position loop is implemented under

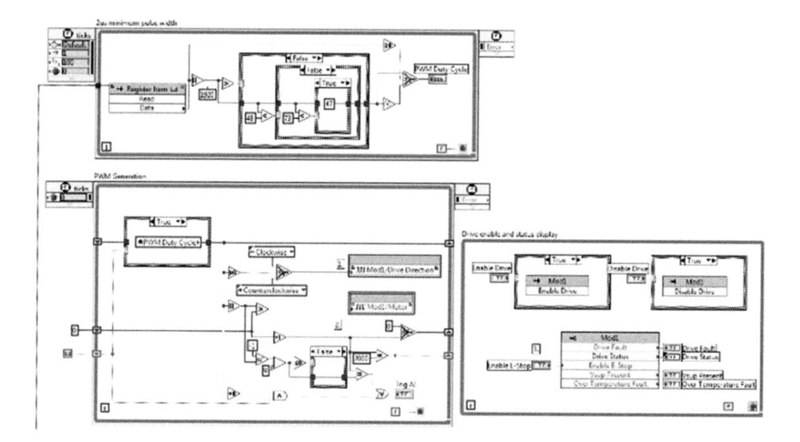

Fig. 4.52 LabVIEW program for generating the PWM signals

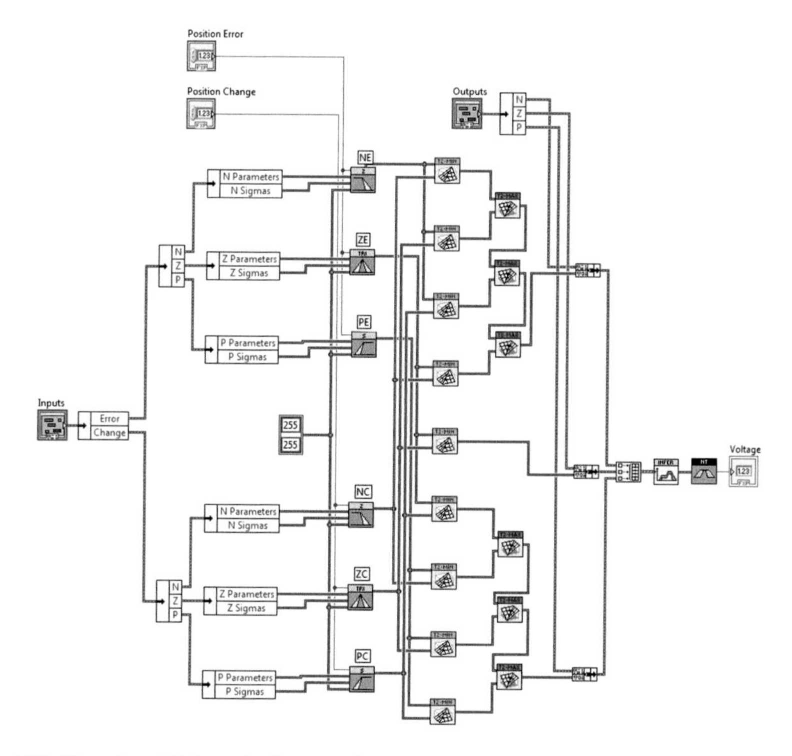

Fig. 4.53 Type-1 and 2 fuzzy logic controller

nonideal isolation conditions and the controller does not tolerate the noise inside the position loop, the precision is degraded dramatically because the complete control loop is corrupted. However, there are some manufacturing procedures that do not require high precision, so PID or fuzzy logic controllers type 1 controllers could be implemented. In fact, the computational cost is lower than fuzzy type 2 and the speed in the entire manufacturing system is increased. On the contrary, fuzzy logic type 2 can deal with noisy signals but the computational cost is higher than fuzzy type 1 and the defuzzification process could be complex if NT method is not selected (see Tables 4.7, 4.8 and 4.9). The following experimental tests were designed to cover different levels of noise and diverse position reference signals in order to capture the performance of each controller. If the controller's performance is known, a correct selection about the kind of controller could be done; the experimental tests were based on the next premises. The CNC micromachine has three DC motors that take over the position in a 3 dimensional plane (*X*, *Y*, and *Z*) and the DC motors have the same characteristics and performance; each position controller used in the CNC micromachine are equal (axis *X*, *Y* and *Z*), so one electric actuator is only studied and presented. Consequently, the controller was developed for *X* dimension (one motor) and replicated to *Y* and *Z* dimensions. Normally, the position control of manufacturing in CNC micromachines can be divided in two trajectories: large and short. By those trajectories, the position command signal can cover the requirements during manufacturing operation. The large position profile moves the actuator through its complete axis and the short trajectory moves slightly the axis from zero to 0.1 mm. An entire set of experimental tests was developed in order to illustrate a complete study about the controller's performance. Different reference position signals were applied and the noise robustness was validated under experimental tests. The first part of the study was done by large position reference signals, which can go from 0 to 15 mm; a noisy signal was added to the control input for confirming the robustness of fuzzy logic controllers and conventional ones. Figure 4.54 depicts the position responses when a source of noise is included into the input control signal and the controller has to cope with it. The results show that conventional and fuzzy controllers have almost the same robustness to the noisy conditions when a large reference of position was applied. Furthermore, the real-time FPGA helps follow the position command signals in deterministic time. Although the fuzzy type 2 performance is diminished due to noisy signals, the performance is good enough for some manufacturing process; in addition, these micromanufacturing machines can be considered as a high performance CNC machines because the position error is very

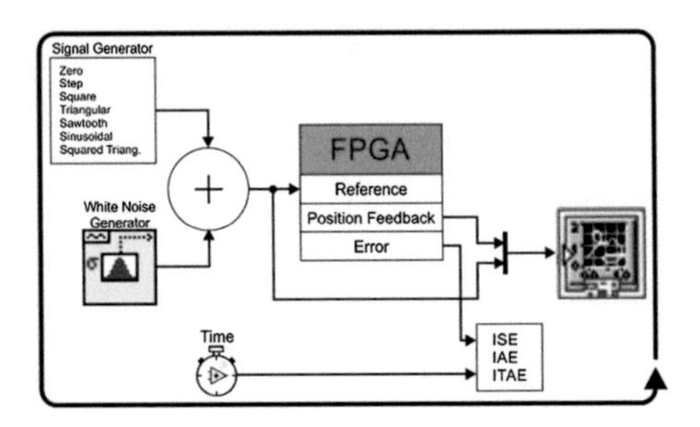

Fig. 4.54 Functional Panel

low. In conclusion, Fig. 4.55 makes available information about the conventional PID and fuzzy controllers; this information gives the option of using PID controllers running in real-time under large position references or without noisy signals. The derivative action in conventional controllers normally affects the performance of the controller when a noisy signal is included [31] but this problem can be decremented when the derivative action is well tuning for a specific system; Thus, different techniques for tuning PID controllers are developed [2] but the complexity of the algorithm increases dramatically and the simplicity of the PID control law disappears. Thus, the proposed fuzzy logic type 2 is attractive because it has a simple topology and achieves excellent position results.

In short position references, the DC actuator has to travel from 0 to 0.1 mm; a noisy signal was also included to evaluate the robustness of the controller. Figure 4.56 shows the results for small position signals when noisy signal is added; in contrast, each controller has different position performance and the fuzzy logic controller type 2 implemented generates excellent performance under short references with noisy signals. The main goal of this test is to validate which controller has the best performance when short position trajectories are applied; short trajectories require high position performance and they are implemented when the work-piece has complex contours; this experimental test demands robustness and precision. The results of fuzzy type 2 show the excellent performance that can be reached when real-time platform is applied. A criterion for selecting, which controller has to be implemented in short position references, is established by indexes of performance, so Table 4.6 presents a complete list of results based on indexes of performance. The indexes studied are the following:

- Integral of Squared Error (ISE), $\int_0^\infty \left[e(t)^2\right] \mathrm{d}t$ (large errors will increased the ISE values)
- Integral of Absolute value of the Error (IAE), $\int_0^\infty \left[e(t)^2\right]$ (ISE will favor small errors)
- Integral of the Absolute value of the Error multiplied by the Time (ITAE), $\int_0^\infty t[e(t)]\mathrm{d}t$ (ITAE penalizes heavily errors that do not appear early in time)
- Transient parameters [Rise time in seconds (t_r), Peak time in seconds (t_p) and Settling Times in seconds (t_s) and the Overshoot expressed as percentage (PO)].

Summing up, the conventional controller, the fuzzy logic controller type 1 and type 2 can reach excellent position response when large command position signals are sent according to Fig. 4.54. Thus, PID and fuzzy logic controllers can be implemented for this kind of position references and they could be a good alternative because they have a simple algorithm that is deployed in real time FPGA; however, in manufacturing systems, short position references are required, so fuzzy logic type 2 has to be considered as the best option under those conditions (see Fig. 4.56). Undoubtedly, real time FPGA allows to complete the controller's algorithm in the time required by deterministic time conditions. This time situations are critical when short position references are needed in manufacturing process and noise shows

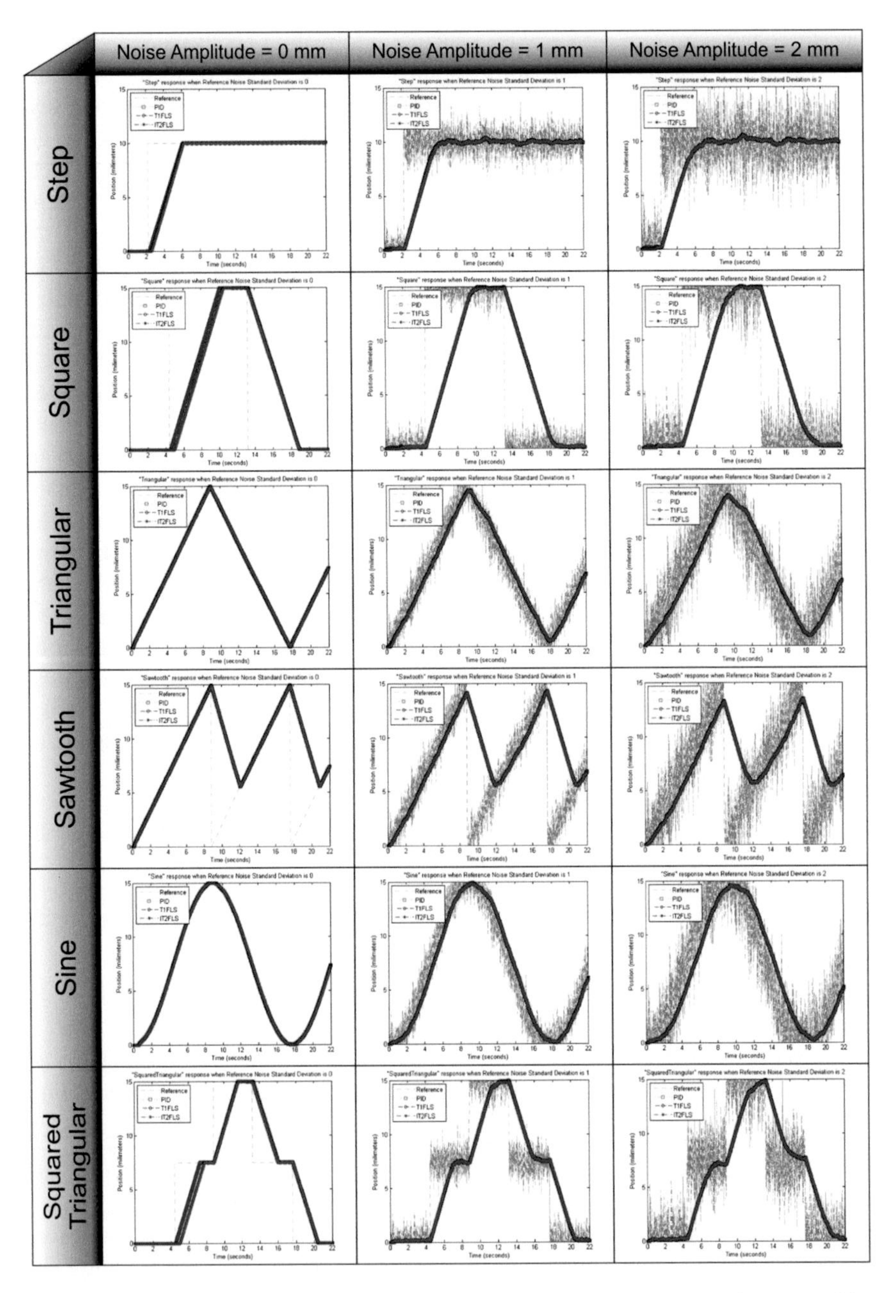

Fig. 4.55 Large travel response with a step response amplitude of 10 mm; the rest position references have an amplitude of 15 mm

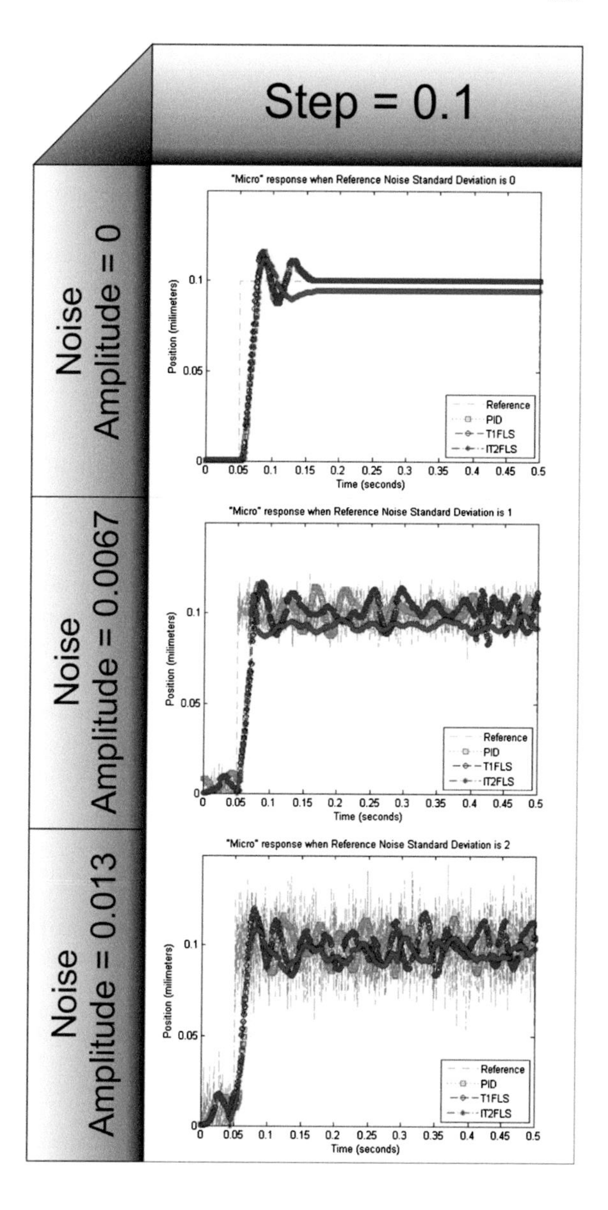

Fig. 4.56 Small position reference (step amplitude of 0.1 mm)

up. Table 4.11, illustrates that the fuzzy logic type 2 has almost the worst ISE, IAE and ITAE values when noise is not presented but it gets the best transient response in the entire set of experiments. When the noise appears, every controller decreases their performance, and the fuzzy type 2 becomes the best option. One of the main drawbacks, in the PID and fuzzy type 1 performances, is created by noisy signals when the amplitude of noise increases, the performance of fuzzy and conventional PID decrease. On the other hand, fuzzy type 2 copes with noisy signals and the number of linguistic rules is kept lower. Hence, the FPGA real time gives good

Table 4.11 Performance parameters in every controller for the small travel response

Noise amplitude	Noise variation SNR (dB)	ISE	IAE	ITAE	t_r (s)	t_p (s)	t_s (s)	OP (%)
PID								
0	∞	0.000123	0.002228	0.00019	0.010038	0.085606	0.093939	15.95
0.006667	23.05	0.000135	0.004843	0.000947	0.046212	∞	∞	15.95
0.013333	17.01	0.000196	0.006918	0.0016	0.049053	∞	∞	16.69
T1FLS								
0	∞	0.000122	0.002183	0.000186	0.009091	0.083523	0.094886	15.95
0.006667	23.05	0.000133	0.004816	0.001021	0.026894	∞	∞	16.69
0.013333	17.01	0.000192	0.007278	0.001747	0.044129	∞	∞	20.92
T2FLS								
0	∞	0.000158	0.004573	0.00082	0.006061	0.086553	0.104924	11.73
0.006667	23.05	0.000143	0.005157	0.001079	0.010038	∞	∞	10.48
0.013333	17.01	0.000165	0.006128	0.00139	0.017045	∞	∞	8.99

results as long as the control law is simple and it does not include a big number of complex calculations. Normally, high performance micromachines require dynamic position responses that have a minimum transitory and stationary error; hence, the amplitude of noise was selected according to the real amplitude of noise. The high performance micromachines do not integrate position sensors that allow limiting the levels of noise [29, 35].

References

1. M. Ramírez, J. Miranda, O. Davila, G. Tello, A. Molina, Reconfigurable didactic microfactory with universal numerical control, in *14th IFAC Symposium on Information Control Problems in, INCOM'2011*, vol. 14, pp. 463–468, 2011
2. S.T. Fleming, D.B. Thomas, FPGA based control for real time systems, in *23rd International Conference on Field Programmable Logic and Applications (FPL)*, 2013. doi:10.1109/FPL.2013.6645610
3. J. Paulo Davim, M.J. Jackson, *Nano and micromachining* (Wiley, New York, 2010), 212 pages
4. O. Iordan, I. Vicol, A.M. Matei, A. Grama, Current state of research in the field on micromilling, http://mech-ing.com/journal/Archive/2010/6/1.mashini/
5. F. Huo, A.-N. Po, Precision countouring control of machine tools. Int. J. Adv. Manuf. Technol. **64**, 319–333 (2013)
6. P. Piljek, Z. Keran, M. Math, Micromachining. Interdisc. Description Complex Syst. **12**(1), 1–27 (2014)
7. E. Kamenar, S. Zelenika, Micropositioning mechatronics system based on FPGA architecture, in *36th International Convention on Information & communication Technology electronics & Microelectronics (MIPRO)*, pp. 125–130, 2013
8. L.A. Zadeh, The concept of a linguistic variable and its application to approximate reasoning —I. Inf. Sci. (Ny). **8**(3), 199–249 (1975)
9. L.A. Zadeh, 1965 J(Zadeh) Fuzzy Sets.pdf. Inf. Control **8**, 338–353 (1965)

10. D. Wu, W.W. Tan, A simplified architecture for type-2 FLSs and its application to nonlinear control, in *IEEE Conference Cybernetics and Intelligent Systems 2004*, vol. 1, pp. 485–490, 2004
11. S. Coupland, R. John, A fast geometric method for defuzzification of type-2 fuzzy sets. IEEE Trans. Fuzzy Syst. **16**(4), 929–941 (2008)
12. Q. Liang, J.M. Mendel, Interval type-2 fuzzy logic systems: theory and design. IEEE Trans. Fuzzy Syst. **8**(5), 535–550 (2000)
13. J.M. Mendel, L. Fellow, R.I. John, F. Liu, S. Member, Interval type-2 fuzzy logic systems made simple. Fuzzy Syst. IEEE Trans. **14**(6), 808–821 (2006)
14. J.M. Mendel, Uncertainty bounds and their use in the design of interval type-2 fuzzy logic systems. IEEE Trans. Fuzzy Syst. **10**(5), 622–639 (2002)
15. D. Wu, J.M. Mendel, On the continuity of type-1 and interval type-2 fuzzy logic systems. IEEE Trans. Fuzzy Syst. **19**(1), 179–192 (2011)
16. L.A. Lucas, T.M. Centeno, M.R. Delgado, General type-2 fuzzy inference systems: analysis, design and computational aspects, in *IEEE International Conference on Fuzzy Systems*, vol. 5, pp. 1–6, 2007
17. K. Duran, H. Bernal, M. Melgarejo, Improved iterative algorithm for computing the generalized centroid of an interval type-2 fuzzy set, in *NAFIPS 2008–2008 Annual Meeting of the North American Fuzzy Information Processing Society*, pp. 1–5, May 2008
18. D. Wu, J.M. Mendel, Enhanced Karnik–Mendel algorithms. IEEE Trans. Fuzzy Syst. **17**(4), 923–934 (2009)
19. M. Nie, W.W. Tan, Towards an efficient type-reduction method for interval type-2 fuzzy logic systems. IEEE Int. Conf. Fuzzy Syst. **2**, 1425–1432 (2008)
20. J.M. Mendel, On a 50 % savings in the computation of the centroid of a symmetrical interval type-2 fuzzy set. Inf. Sci. (Ny) **172**(3–4), 417–430 (2005)
21. J.M. Mendel, L. Fellow, X. Liu, simplified interval type-2 fuzzy logic systems. IEEE Trans. Fuzzy Syst. **21**(6), 1056–1069 (2013)
22. A. Tellez-Velazquez, H. Molina-Lozano, L.A. Villa-Vargas, The Tellez-Molina-Villa Algorithm, in *2012 Annual Meeting of the North American Fuzzy Information Processing Society*, pp. 1–6, Aug 2012. doi:10.1109/NAFIPS.2012.6291017
23. H.-J. Wu, Y.-L. Su, S.-J. Lee, A fast method for computing the centroid of a type-2 fuzzy set. IEEE Trans. Syst. Man Cybern. B Cybern. **42**(3), 764–777 (2012)
24. Available:http://motionsystems.pi-usa.us/item/tioning-actuators-stages-linear-translation-stages/tion-micro-translation-stages-with-ballscrew-drive/m-111-2dg
25. T. Dam, P-R. Ouyang, Position domain contour tracking with cross-coupled control, in *IEEE International Symposium on Industrial Electronics (ISIE)*, pp. 1303–1308, 2012
26. Y. Koren, C.C. Lo, Advanced controllers for feed drives. CIRP Ann.-Manuf. Appl. Int. Conf. **41**(2), 689–698 (1992)
27. O. Castillo, P. Melin, *Type-2 Fuzzy Logic: Theory and Applications* (Springer Science, Berlin). ISBN: 978-3-540-76283-6
28. X. Ling, Q. Li, T. Wang, J. Dong, Z. Tang, Y. Ding, Research and decision of motion controller for CNC based on fuzzy PID algorithm with feedforward control, in *International Conference on Control, Automation and Systems Engineering (CASE)*, pp. 1–4, 2011
29. NI: NI 9505 (FPGA Interface). [Online]. Available: http://zone.ni.com/reference/en-XX/help/370984R-01/target4devicehelp/9505_io_reference/
30. LabVIEW FPGA website: www.ni.com, revised 2014
31. K. Ogata, *Discrete-Time Control Systems*, vol. 2. (Prentice Hall, Englewood Cliffs, NJ, 1995)
32. P. An-Hua, Gra-based approach to PID parameter tuning for closed-loop servo systems, in *Control and Decision Conference (CCDC) 25th Chinese*, pp. 1628–1633, 2013
33. A.Y. Jaen-Cuellar, R.J. Romero-Troncoso, L. Morales-Velazquez, R.A. Osornio- Rios, PID-controller tuning optimization with genetic algorithms in servo systems. Int. J. Adv. Robot. Syst. (2013). doi:10.5772/56697
34. cRIO Manual: (http://www.ni.com/pdf/manuals/374126e.pdf)

35. NI: NI 9505 PWM Generation-Improved. [Online]. Available: http://www.ni.com/example/30177/en/
36. J.M. Mendel, Type-2 fuzzy sets and systems: an overview. IEEE Comput. Intell. Mag. **2**(1), 20–29 (2007)
37. J. Mendel, F. Liu, Super-exponential convergence of the Karnik–Mendel algorithms used for type-reduction in interval type-2 fuzzy logic systems, in *IEEE International Conference on Fuzzy Systems,* Vanco

Index

© Springer International Publishing Switzerland 2016
P. Ponce-Cruz et al., *Fuzzy Logic Type 1 and Type 2 Based on LabVIEW™ FPGA*, Studies in Fuzziness and Soft Computing,
DOI 10.1007/978-3-319-26656-5

Printed in the United States
By Bookmasters